Richard Legay

The Cultural Histories of Radio Luxembourg and Europe n°1

Broadcasting Across Europe

palgrave
macmillan

Richard Legay
Arnold-Bergstraesser-Institut
Freiburg, Baden-Württemberg
Germany

ISBN 978-3-031-46249-8 ISBN 978-3-031-46250-4 (eBook)
https://doi.org/10.1007/978-3-031-46250-4

This Palgrave Macmillan imprint is published by the registered company Springer Nature Switzerland AG
The registered company address is: Gewerbestrasse 11, 6330 Cham, Switzerland

Paper in this product is recyclable.

The Cultural Histories of Radio Luxembourg and Europe n°1

Acknowledgements

This book was inspired by my doctoral dissertation, written under the supervision of Andreas Fickers, to whom I would like to express my gratitude and appreciation for his guidance and mentoring all these years. I would like to extend my thanks for their precious support to Alec Badenoch, Marie Cronqvist, Carolyn Birdsall, Hugh Chignell, Valérie Schafer, Hans-Ulrich Wagner, and Paul Lesch.

I am also grateful for their help to the colleagues at the Centre for Contemporary & Digital History, the University of Saarland, and Bournemouth University. This work would not have been possible without the members of the International Association for Media & History, and of the Entangled Media Histories and Popkult60 networks, who were unfailingly helpful. Finally, I would like to acknowledge the unwavering support of my partner, my friends, and my relatives, who have had my back ever since I embarked on this academic journey.

Acknowledgements

Contents

List of Figures

CHAPTER 1

Introduction

Both Europe n°1 and Radio Luxembourg, commercial radio stations that have been broadcasting for decades now, had, in the 1960s, a wide community of listeners all over Europe and an undeniable popularity. The former broadcast in French, while the latter had services in multiple languages, including French and English, meaning they had an important listenership in many countries. They were even involved with popular magazines, such as *Fabulous 208* and *Salut les Copains* for which sales were rather successful. Some of their programmes and hosts became iconic and are still remembered, today, by former listeners. As such, the commercial stations had a non-negligible place within the wider European broadcasting and media landscape, alongside public broadcasters which had a prominent role.

The Cultural Histories of Radio Luxembourg and Europe n°1 sets out to uncover the ways these stations worked across media and countries to get into this central position of popular broadcasters in Europe. To do so, it postulates the existence of an overarching *dispositif* that enabled Radio and Europe n°1 to operate on various dimensions (sound, space, and intermediality) and differentiated them from other broadcasters. Answering current calls in media history and radio studies, most notably from the Entangled Media Histories network, this book stands out by embracing a transnational and intermedial lens of study to approach the radio stations at hand. The nature and history of commercial

R. Legay, *The Cultural Histories of Radio Luxembourg and Europe n°1*,
https://doi.org/10.1007/978-3-031-46250-4_1

radio stations, such as Radio Luxembourg and Europe n°1, make them quintessentially transnational actors, and prone to work with other media, such as magazines. Therefore, this is a history of radio that follows Kate Lacey's open definition of the medium, a word which is 'called upon to describe any number of different things',[1] by notably relying on the analysis of a wide range of historical sources, including institutional archives, magazines, historical recordings, and maps.

This book studies Europe n°1 and Radio Luxembourg, its French and English services, both as separate institutions and as a whole, coherent group, due to their proximity as commercial radio stations. It inquires a multitude of different sources: their radio programmes, their magazines, their institutional archives, etc.—to deconstruct the *dispositif* of commercial radio stations in Western Europe in the 1960s. This work, therefore, embarks on the analysis of the various components shaping this transnational *dispositif*. This includes looking into the institutional archives, analysing magazines published by the stations, and digging into the soundscape of commercial radio stations in the 1960s, through audio sources. These different elements are studied throughout the book and constitute key sections of this work. Their successful analysis brings answers regarding the hypothesis created around the development of a transnational *dispositif*.

The research behind this book finds its roots in the field of media history, more precisely in radio history. It is now a well-established area, with national and international scholarly networks, almost as flourishing as the fields of film or television history. This monograph draws on the existing literature on radio history and branches out towards other fields of history and social sciences to answer questions raised throughout the research.[2] To do so, *The Cultural Histories of Radio Luxembourg and Europe n°1* relies on a wide range of scholarly literature. To give a sense of this, the following pages will give insights into the current state of research regarding (commercial) radio history and then will focus on a

[1] Kate Lacey, "Up in the air? The matter of radio studies", *The Radio Journal: International Studies in Broadcast & Audio Media* (16:2, 2018), pp. 109–126.

[2] As argued by Kate Lacey, "this singular word, radio, is called upon to describe any number of different things - material, virtual, institutional, aesthetic, experiential", which reinforces the necessity for this book to avoid a fixed single-medium perspective. Kate Lacey, "Up in the air? The matter of radio studies", *Radio Journal: International Studies in Broadcast & Audio Media* (16:2, 2018), pp. 109–126.

series of concepts central to the research undertaken for this book. They have been used in both radio history and other disciplines and will help situate this book into a larger scholarly landscape.

Radio History. There is no need, anymore, to justify, and apologise for, a work focusing on radio as a single medium, as Kate Lacey pointed out.[3] This comment, however, reveals the long and difficult path that radio historians have undertaken to make sure their specialty would no longer be overlooked. The 'curse' of radio history has often been to be melted into the wider term of media or broadcasting history, where it generally played a minor role.[4] A large range of publications, however, have since helped radio studies, and, by extension, radio history, win their spurs. Thankfully, one strength of the field of radio history has been the permeability between scholars and practitioners, meaning that books that were published for radio studies become particularly useful tools in the hands of a historian. Two handbooks for radio studies, one in French, and the other one in English, respectively written by Frédéric Antoine[5] and Hugh Chignell,[6] have been extremely useful to define and properly use concepts and definitions throughout this work and were often used jointly in order to tackle the difficulties created by a multilingual research context. These books also highlight the importance of properly grasping relevant source material, especially audio documents, to analyse them to their full extent. In this regard, the chapter published by Céline Loriou in *Le Goût de l'archive à l'ère numérique* is a fascinating reconstruction of the contemporary archival experience of working with historical radio

[3] Kate Lacey, "Ten Years of Radio Studies: The Very Idea", *The Radio Journal: International Studies in Broadcast & Audio Media* (6:1, 2008), pp. 21–32.

[4] "Despite its ubiquitous presence in American life for over half a century, radio is a medium whose political and cultural power and influence are not yet reflected in American historiography, American studies, work on American race relations, or studies of the media and popular culture". See Barbara D. Savage, *Broadcasting Freedom* (Chapel Hill: The University of North Carolina Press, 1999), pp. 5–6. For a long time, courses on radio were applied, practical, and only recently, radio 'studies' stopped being marginalised. See Michael C. Keith, "The Long Road to Radio Studies", *Journal of Broadcasting & Electronic Media* (51:3, 2007), pp. 530–536.

[5] Frédéric Antoine, *Analyser la radio. Méthodes et mises en pratique* (Louvain-la-Neuve: De Boeck, 2016).

[6] Hugh Chignell, *Key Concepts in Radio Studies* (London: Sage, 2009).

sources.[7] Radio history is not only the study of radio and broadcasting institutions. Many other aspects and research perspectives can enter this field, and have a crucial role in the understanding of both Radio Luxembourg and Europe n°1. It is also, among other things, the history of societies, cultures, practices and rituals, and technologies.[8] Radio studies show a wide range of interests that can enrich the understanding of *how* these commercial radio stations were consumed and listened to by their audiences. Many of the most popular programmes broadcasted by commercial radio stations targeted a young listenership and were on air at the end of the day and at night, which is, in many ways, a unique social moment. In this regard, the research done by Marine Beccarelli on night-time radio in France is a powerful example of historical inquiry into the connections between listeners and radio stations.[9] The historical study of Radio Luxembourg and Europe n°1 is closely linked to material culture and the history of technology. The works of Elvina Fesneau and of Andreas Fickers regarding the history of the transistor set[10] are particularly useful to highlight this aspect of the radio experience, as it became a predominant item all around Europe in the 1960s, especially for younger generations. Other works contribute to a better understanding of the technological aspects of radio. Among them, Michael Bull's *Sounding Out the City*[11] makes for a good addition to Fesneau's research as it

[7] Céline Loriou, "Faire de l'histoire, un casque sur les oreilles: le goût de l'archive radiophonique", in F. Clavert & C. Muller (eds.), *Le Goût de l'archive à l'ère numérique* (first published online in 2018, [http://www.gout-numerique.net/table-of-contents/faire-de-lhistoire-un-casque-sur-les-oreilles-le-gout-de-larchive-radiophonique], last consulted 17/03/2023).

[8] In this respect, media studies offer many great insights to better understand radio, its impact and functioning. Among the large body of literature on this topic, see, for instance, Paddy Scannell, *Media and Communication* (London: Sage, 2007); and Jostein Gripsrud, *Understanding Media Culture* (London: Arnold, 2002).

[9] See Marine Beccarelli, *La Nuit du bout des ondes* (Paris: INA Editions, 2014), and Marine Beccarelli, *Micros de nuit: histoire de la radio nocturne en France, 1945-2012* (Rennes: Presses Universitaires de Rennes, 2021).

[10] Elvina Fesneau, *Le Transistor à la conquête de la France. La radio nomade (1954-1970)* (Paris: INA Editions, 2014) & Andreas Fickers, *Der 'Transistor' als technisches und kulturelles Phänomen: Die Transistorisierung der Radio- und Fernsehempfänger in der deutschen Rundfunkindustrie, 1955-1965* (Bassum: Verlag für Geschichte der Naturwissenschaften und Technik, 1998).

[11] Michael Bull, *Sounding Out the City*, 2nd ed. (Oxford: Berg, 2006, 1st edition 2000).

focuses on the personal experience of listening and the connections to material culture.

The historical literature on commercial radio in Europe itself is light. And, as if to mirror their French name—*radios périphériques*—they are still somewhat on the periphery of the main historiography. There is indeed a disbalance between the stations studied in this book, and it is easily illustrated by this fact: until Denis Maréchal's book in 2023, there had only been one monograph published about the history of Europe n°1.[12] Despite the lack of scholarly literature, there is a large amount of 'grey literature' written about the station. For instance, an illustrated dictionary about Europe n°1[13] was published, as well as several books dealing specifically with one of the station's most iconic shows: *Salut les Copains*.[14] Obviously, they do not provide scientific analysis, but are rather helpful for contextualisation, notably as they are usually abundant with pictures. On the other hand, while grey literature exists about Radio Luxembourg,[15] the station has also been under more scrutiny from radio scholars, particularly the French service. It seems that there are two waves when it comes to the historiography of Radio Luxembourg. First, a trilogy of books by Denis Maréchal and David Dominguez Muller, published in 1994,[16] 2007,[17] and 2010[18] represents a milestone in the understanding of the station's history, especially its creation and early days. Second, the

[12] See Denis Maréchal, *Europe 1: De la singularité au déclin (1955–2022)* (Lormont: Bord de l'Eau, 2023) & Luc Bernard, *Europe 1: La Grande histoire dans une grande radio* (Paris: Centurion, 1990).

[13] D. Olivennes, F. Ferrand & B. Labous, *Europe 1: Le Dictionnaire amoureux illustré* (Paris: Plon, 2015).

[14] Christophe Quillien, *Nos Années Salut les Copains 1959-1976* (Paris: Flammarion, 2009) & Gilbert Jouin, *Salut les Copains* (Paris: Ipanema Éditions, 2012).

[15] Among the several possible examples, see Jean-Pierre Defrain & Jacques Boutelet, *RTL. 40 ans ensemble* (Paris: Calmann-Lévy, 2006).

[16] Denis Maréchal, *Radio-Luxembourg 1933-1993, un média au coeur de l'Europe*, (Nancy: Presses Universitaires de Nancy, 1994).

[17] David Dominguez Muller, *Radio-Luxembourg: Histoire d'un média privé d'envergure européenne*, (Paris: L'Harmattan, 2007).

[18] Denis Maréchal, *RTL. Histoire d'une radio populaire: De Radio Luxembourg à RTL.fr*, (Paris: Nouveau Monde Éditions, 2010).

work done by Anna Jehle[19] and Katja Berg,[20] who represent the 'new wave' of historical inquiry. Their monographs are a great resource as they are the most up-to-date publications on the subject. They even co-published an article that looked at Radio Luxembourg as a transnational actor, through a Franco-German focus, already showing the potential of a transnational inquiry into the history of this commercial radio station.[21] This historiographical 'new wave' on Radio Luxembourg proves the relevance of a transnational perspective while revealing, at the same time, that there is still room for inquiry. Most importantly, it shows the necessity to make commercial radio stations—as a whole—central in the discussion, which can only be done through a transnational angle, which has to be pushed forward.

It is crucial here to mention that commercial radio stations are rarely studied together, but rather individually, as it was just mentioned, or as the 'others', *i.e.* what is not public broadcast, something especially noticeable when looking at France. There is, however, potential to embrace them as a whole, to fully understand their impact on the whole European media landscape. They do appear, at times, in more general history of media or radio in France especially, often as the 'alterity', the *postes périphériques*, those that do not belong—at least not fully—to the national historiography.[22] A similar dynamic can be seen for Radio Luxembourg

[19] Anna Jehle, *Welle der Konsumgesellschaft: Radio Luxembourg in Frankreich, 1945-1975*, (Göttingen: Wallstein Verlag, 2018).

[20] Katja Berg, *Grenzenlose Unterhaltung: Radio Luxemburg in der Bundesrepublik, 1957–1980* (Göttingen: Wallstein Verlag, 2021).

[21] Katja Berg & Anna Jehle, "'Through the Air to Anywhere': Radio Luxembourg—A Transnational Broadcasting Station?" in Christoph Classen (ed.), *Transnational Broadcasting in Europe 1945-1990* (Frankfurt am Main: Peter Lang, 2016), pp. 23–44. The PhD dissertation written by Jennifer Spohrer is also of great interest to reveal the transnationalism of the early days of Radio Luxembourg. See Jennifer Spohrer, "Ruling the Airwaves: Radio Luxembourg and the Origins of European National Broadcasting, 1929–1950", PhD diss. (Columbia University, 2008). In a similar way, an article written by Gerald Newton on the history of Radio Luxembourg during the Second World War is also significant for the station's transnational history. Gerald Newton, "Radio Luxembourg in Peace and War", *German Life and Letters* (66:1, 2013), pp. 55–76.

[22] Jean-Jacques Cheval devotes, for example, a handful of pages to commercial radio stations, including Radio Luxembourg and Europe n°1 in his general book on French Radio. See Jean-Jacques Cheval, *Les Radios en France: Histoire, état et enjeux* (Rennes: Apogée, 1997).

in the British context, where it is often presented as an outsider and challenger to State monopoly,[23] especially in its early days, which have been studied in more detail.[24] There are a few reasons that might explain this marginalisation. One being the access to their archives, which can be challenging.[25] One can also make the hypothesis that these radio stations do not have the same reputation as their public counterparts, which are seen as more intellectual, and more serious—in other words—more prestigious, than commercial enterprises, hence why historians and scholars have not studied them with the same interest until recently.

Transnational History has had an essential impact on the research conducted for this book. Indeed, radio history has been, for a long time, dominated by studies framed by national borders and state-owned institutions. However, over the last decade or so, many works in broadcasting history overall have embraced a 'transnational turn'. It echoes strongly with this book as commercial radio stations are quintessentially transnational actors.[26] Transnational history can be defined as a tradition[27] that seeks to analyse the complex flows of ideas, goods, people, etc. across national borders. It does not reject the importance of national entities, but rather the idea that nation-states are a *natural* framework for historical analysis, and problematises this analytical frame.[28] This history of 'circulations and connections'[29] appears especially useful for broadcasting history, a field in which cross-border circulations have always played a

[23] John Jenks, "Commercial temptation: Cross-border radio and the comparative transformation of public service broadcast policy in Britain, South Africa and India, 1930-67", *Radio Journal: International Studies in Broadcast & Audio Media* (15:1, 2017), pp. 125–141.

[24] See, for example, Seán Street, *Crossing the Ether: Pre-War Public Service Radio and Commercial Competition, 1922–1945* (Eastleigh: John Libbey Publishing, 2006); and Richard Nichols, *Radio Luxembourg: The Station of the Stars* (London: W.H. Allen, 1983).

[25] Richard Legay, 'Studying 1960s commercial radio', in K. McDonald & H. Chignell (eds.), *Bloomsbury Handbook of Radio* (London: Bloomsbury, 2023), pp. 367–382.

[26] Andreas Fickers, "Radio", in Akira Iriye & Jean-Yves Saunier (eds.), *Palgrave Dictionary of Transnational History* (Basingtoke: Palgrave Macmillan, 2009), pp. 870–873.

[27] A concise summary on the historiography of this tradition can be found in Akira Iriye, *Global and Transnational History* (Basingstoke: Palgrave Macmillan, 2013), pp. 36–68.

[28] Marie Cronqvist & Christoph Hilgert, "Entangled Media Histories. The Value of Transnational and Transmedial Approaches in Media Historiography", *Media History* (23:1, 2017), pp. 130–141.

[29] Iriye & Saunier, *Palgrave Dictionary of Transnational History*, pp. xvii–xx.

crucial role. Furthermore, the entry for 'broadcasting' in the *Palgrave Dictionary of Transnational History* states that transnationalism is an inherent quality of broadcasting due to the very nature of radio waves.[30] Transnational broadcasting is not solely the communication via radio and television across borders, but it also involves the 'basic political, technical, economic and social conditions and contexts',[31] meaning that it goes further than the transnational technical act of cross-border broadcasting. Christoph Classen also explains that his definition of transnational broadcasting includes programmes aimed at a foreign country, but also those which simply transcend borders.[32] This is crucial for Radio Luxembourg and Europe n°1, for they appropriated a large transnational broadcasting space, due to having listeners in many countries, which is a key topic in this book. A few other publications should be mentioned as they strongly impacted the transnational dimension of the research undertaken in this book. Among them, Andreas Fickers and Pascal Griset published *Communicating Europe*,[33] the fruit of a long collaboration on the transnational history of communication technology in Europe. A special issue of the *Historical Journal of Film, Radio and Television* dedicated to transnational radio history offered a series of case studies covering a wide range of regions, perspectives, and actors.[34] Europe has often been at the centre of transnational broadcasting history research, notably in Suzanne Lommers' *Europe—On Air*,[35] and it appears that historians found the Cold War a particularly rich period to articulate research on transnational broadcasting.[36]

[30] Andreas Fickers, "Broadcasting", in Iriye & Saunier, *Palgrave Dictionary of Transnational History*, pp. 106–108.

[31] Christoph Classen, "Introduction" in C. Classen (ed.), *Transnational Broadcasting in Europe 1945-1990* (Frankfurt am Main: Peter Lang, 2016), pp. 7–22.

[32] Christoph Classen (ed.), *Transnational Broadcasting in Europe 1945-1990* (Frankfurt am Main: Peter Lang, 2016).

[33] Andreas Fickers & Pascal Griset, *Communicating Europe: Technologies, Information, Events* (Basingstoke: Palgrave Macmillan, 2019).

[34] See T. Hochscherf, R. Legay & H. Wagner, "Radio Beyond Boundaries: An Introduction", *Historical Journal of Film, Radio and Television* (39:3, 2019), pp. 431–438.

[35] Suzanne Lommers, *Europe—On Air. Interwar Projects for Radio Broadcasting* (Amsterdam: Amsterdam University Press, 2012).

[36] See for example A. Badenoch, A. Fickers & C. Henrich-Franke (eds.), *Airy Curtains in the European Ether: Broadcasting and the Cold War* (Baden-Baden: Nomos, 2013), & Emily Oliver, "A Voice for East Germany: Developing the BBC German Service's East

Sound Studies have also been particularly insightful in this book. Indeed, it regularly uses sound as source material from a historical perspective.[37] More precisely, recordings of radio programmes from the long 1960s were analysed at length, and in all their sonic richness. This is where sound studies come in handy to fully exploit those archives. Among this vast and rich field, the concept of soundscape was also particularly useful. In addition to R. Murray Schafer's original work,[38] recent historical uses of the concept, particularly Carolyn Birdsall's *Nazi Soundscapes*,[39] should be mentioned, notably they have helped legitimise sound as an object of historical study. When it comes to radio, the soundscape, as Chignell explains, is 'more than just background noise, it is itself the bearer of meaning',[40] which means that every audio element can be analysed as well. On some occasions, some of the recordings, made by amateurs who recorded their radio sets, might even tell the historian about the soundscape in which radio was listened to, as some exterior sonic elements can be present (*e.g.* someone tuning in to another station). As mentioned, sound studies are a rich field and there are multiple publications that offer reflections on the progress and the challenges of the field.[41] More specifically, these volumes include insights into crucial elements, such as the history of the radio dial and the concept of imagined community, which are both used later in this book. Sound history naturally finds some roots in the history of the senses, which can be dated back to—at least—the interwar period and the works of Lucien Febvre.[42]

Zone Programme", *Historical Journal of Film, Radio and Television* (39:3, 2019), pp. 568–583.

[37] Building notably on *Klang als Geschichtsmedium*, which displayed several examples of historical studies using sound as source material, mostly in relation to music. Anna Langenbruch (ed.), *Klang als Geschichtsmedium: Perspektiven für eine auditive Geschichtsschreibung* (Bielefeld: Transcript, 2018).

[38] R. Murray Schafer, *The Soundscape: Our Sonic Environment and the Tuning of the World*, 2nd ed. (New York: Rochester, 1994), p. 8.

[39] Carolyn Birdsall, *Nazi Soundscapes: Sound, Technology and Urban Spaces in Germany, 1933-1945* (Amsterdam: Amsterdam University Press, 2012).

[40] Chignell, *Key Concepts in Radio Studies*, p. 105.

[41] See notably Trevor Pinch & Karin Bijsterveld (eds.), *The Oxford Handbook of Sound Studies* (Oxford: Oxford University Press, 2012), and Jonathan Sterne (ed.), *The Sound Studies Reader* (London: Routledge, 2012).

[42] Lucien Febvre, "Psychologie et histoire" (1938) & "Comment reconstituer la vie affective d'autrefois? La sensibilité et l'histoire" (1941), both reprinted in Lucien Febvre,

Unfortunately, according to David Suisman, this idea found an echo only in the 1980s and the 'pioneering work of Alain Corbin on smell, sound, and touch in the French social imagination,[43] did the potential of Febvre's work start to become manifest'.[44] In the late 1990s, Susan J. Douglas was also disappointed by the lack of sonic scholarship which led to the lack of proper acknowledgement of 'particular qualities and power of sound, and how these have shaped the power of radio'.[45] Hopefully, the analysis of the 'sound' of commercial radio stations conducted in this book sheds some light on the connections between the power of sound and the power of radio.

Being **commercial radio** is what connects the different stations studied in this book. It can be characterised as one of the four categories of radio (public, state-run, and community),[46] and is defined by its specific goal, which is, to put simply, to make a profit. It appears as a defining trait of the topic at hand in this book, for it is the primary purpose of the *dispositif* of commercial radio stations, and, as such, it is what makes them stand out in the European broadcasting landscape. Commercial radio finds its roots in the United States and in Europe, with stations such as Radio Normandy and Radio Luxembourg. This category of radio has to be understood in relation to advertising, the 'sale of airtime to businesses'.[47] On one hand, advertising brings large amounts of money to the stations, but, on the other hand, adverts are, at times, perceived by listeners as negative and intrusive, breaking the rhythm of radio programming. Moreover, advertisers favour predictable and safe formats, limiting, de facto, innovation. It is interesting to note that, according to Hilmes,

A New Kind of History: From the Writings of Febvre (New York: Harper and Row, 1973, translation by K. Folca).

43 Suisman refers here to a series of monographs published by Corbin over the years. For sound, see Alain Corbin, *Les Cloches de la terre: paysage sonore et culture sensible dans les campagnes au XIXe siècle* (Paris: Flammarion, 1994); while for the history of smell, see Alain Corbin, *Le Miasme et la jonquille: l'odorat et l'imaginaire social, XVIIIe-XIXe siècles* (Paris: Flammarion, 1982).

44 David Suisman, "Introduction: Thinking Historically about Sound and Sense", in D. Suisman & S. Strasser (eds.), *Sound in the Age of Mechanical Reproduction* (Philadelphia: University of Pennsylvania Press, 2010), pp. 1–12.

45 Susan J. Douglas, *Listening In: Radio and the American Imagination* (New York: Times Books, 1999), p. 7.

46 Chignell, *Key Concepts in Radio Studies*, p. 114.

47 ibid, p. 111.

the history of radio, and the history of radio advertising agencies, have been evolving alongside each other since the beginning.[48] Historical literature about commercial radio in Europe is inherently linked to the history of Radio Luxembourg and other *radios périphériques*. American broadcasting history, where commercial radio played a crucial role, is also particularly insightful. Derek W. Vaillant's *Across the Waves*[49] and Michele Hilmes' *Network Nations*[50] are especially relevant monographs as they specifically cover both sides of the Atlantic. A more national example is *Stay Tuned: A History of American Broadcasting*,[51] first published in 1978, and now in its third edition. It is particularly helpful because it reflects historiographical evolutions in the field of broadcasting history, and gives a large role to the commercial dimension in a national setting, something obviously absent in the European context.

The concept of **'entanglements'** in media history was coined by the international research network Entangled Media Histories, commonly known as EMHIS, who develops collaboration around transnational and transmedial approaches in the field. When working on commercial radio stations in the 1960s, it quickly became apparent that the boundaries with other media, particularly magazines, were often blurry, and the need for such a conceptual tool that would help define these connections arose. This concept seeks to address the 'blind spots' in media history, especially where different media are influencing one another, something this book is set to do as well. Entanglements are defined as 'intended or unintended, obvious or hidden, structured or chaotic interrelation(s) in space, knowledge or time',[52] that can take many forms (*i.e.* reciprocal, symmetrical), which is partly why it was chosen to work with it. The publication of the article by Hilgert and Cronqvist was followed by commentaries from two prominent scholars of media history. Bridget Griffen-Foley and

[48] Michele Hilmes, *Radio Voices: American Broadcasting, 1922-1952* (Minneapolis: University of Minnesota, 1997), p. 114.

[49] Derek W. Vaillant, *Across the waves: How the United States and France shaped the International Age of Radio* (Champaign: University of Illinois, 2017).

[50] Michele Hilmes, *Network Nations: A Transnational History of British and American Broadcasting* (London: Routledge, 2012).

[51] Christopher H. Sterling & John Michael Kittross, *Stay Tuned: A History of American Broadcasting*, 3rd ed. (Mahwah & London: Lawrence Erlbaum Associates, 2002).

[52] Cronqvist & Hilgert, "Entangled Media Histories", pp. 130–141.

Michele Hilmes presented how the concept could be articulated in relation to their own research and within their academic backgrounds in Australia[53] and the United States.[54] These commentaries are particularly insightful to show potential applications of the concept outside of the research group.[55] In 2020, members of EMHIS contributed to an edited volume on European Communication History,[56] and a special issue of *Media History* was dedicated to 'entangled media histories' and offered concrete applications of the concept.[57] Both publications are useful additions to the more theoretical original article, which is a reference point for the intermedial aspect of the research undertaken in this book.

Therefore, the concept of 'entangled media histories' is the most appropriate tool to include transnational approaches with transmedial ones, and go towards an integrated media history.[58] This notably means relying on the history of magazines. Indeed, periodicals published in close collaboration with commercial radio stations, such as *Salut les Copains* and *Fabulous 208*, are extensively used in this book. To do so, the work of Chris Tinker has been particularly helpful, as he has made issues of this magazine central to his research.[59] His work also includes a comparative approach to the question of 'masculinities' through the study of

[53] Bridget Griffen-Foley, "Entangled Media Histories: A Response", *Media History* (23:1, 2017), pp. 145–147.

[54] Michele Hilmes, "Entangled Media Histories: A Response", *Media History* (23:1, 2017), pp. 142–144.

[55] They also led to another commentary from the authors of the original article. Marie Cronqvist & Christoph Hilgert, "Entangled Media Histories: Response to the Responses", *Media History* (23:1, 2017), pp. 148–149.

[56] H.-U. Wagner, H. Chignell, M. Cronqvist, C. Hilgert and K. Skoog, "Media after 1945: Continuities and New Beginnings", in K. Arnold, P. Preston & S. Kinnebrock (eds.), *The Handbook of European Communication History* (London: Wiley, 2020), pp. 189–204.

[57] Christoph Hilgert, Marie Cronqvist & Hugh Chignell, "Tracing Entanglements in Media History", *Media History* (26:1, 2020), pp. 1–5.

[58] Sîan Nicholas, "Media History or Media Histories? Re-addressing the history of mass-media in inter-war Britain", *Media History* (18:3-4, 2012), pp. 379–394.

[59] Chris Tinker, *Mixed Messages: Youth Magazines Discourse and Sociocultural Shifts in Salut les Copains (1962–1976)* (Oxford, New York: Peter Lang, 2010). Chris Tinker, "Shaping Youth in Salut les Copains", *Modern & Contemporary France* (15:3, 2007), pp. 293–308.

Salut les Copains and *Fabulous*,[60] which shows the similarities between the two magazines, and the relevance of studying them together as in this book. Moreover, Claire Blandin has inquired into the relationship between both *Salut les Copains*, the magazine, and *Salut les Copains*, the radio programme, as a youth cultural product.[61] These two authors have paved the way to further an inquiry into the entanglements between radio stations, radio programmes, and youth magazines.

The concept of **dispositif**,[62] in a few words, is a term used to refer to mechanisms and structures—of various nature—used to maintain a form of power within the social body. It first appears—according to film historian Frank Kessler—[63] within cinema studies and the work of Jean-Louis Baudry,[64] and was also central in the work of French philosopher and social theorist Michel Foucault. The first glimpses of the concept are indeed found in Foucault's inaugural lecture at the prestigious *Collège de France* in Paris in 1970; however, he did not mention it directly, rather hinting at something close to it. The lecture was slightly amended and later published in *L'Ordre du Discours*, and, in addition to interviews in the following years, constitutes founding elements for the definition of the concept. The *dispositif* then found its way into social sciences and was later reused, further developed, and employed by a variety of scholars. Among them, are French philosopher Gilles Deleuze and Italian philosopher Giorgio Agamben. Regarding Foucault's take on the concept,

[60] Chris Tinker, "Mixed Masculinities in 1960s British and French Youth Magazines", *The Journal of Popular Culture* (7:1, 2014), pp. 84–108. Chris Tinker, "Shaping 1960s youth in Britain and France: Fabulous and Salut les Copains", *International Journal of Cultural Studies* (14:6, 2011), pp. 641–657.

[61] Claire Blandin, "Radio et magazine: une offre plurimédia pour les jeunes des Sixties", *Le Temps des Médias* (21:2, 2013), pp. 134–142.

[62] There are debates surrounding the translation of the French term *dispositif* in English. Indeed, some authors suggest 'dispositive' or 'apparatus', nevertheless they do not convey the entire original meaning. It was therefore chosen to keep the French word *dispositif*, to make sure the concept retains all its original significance. Etymologically speaking, Valérie Larroche traces the early definition of the word *dispositif* in French to notions of organisation and arrangements, with a strong legal component, while the links to technique appeared later and now have precedence. See Valérie Larroche, *The Dispositif: A Concept for Information and Communication Sciences* (London: ISTE, 2019), pp. xiv–xv.

[63] Frank Kessler, "Notes on *Dispositif*", published online [http://www.frankkessler.nl/wp-content/uploads/2010/05/Dispositif-Notes.pdf], 2007 (last accessed 12/08/2020).

[64] Jean-Louis Baudry, *L'Effet cinéma* (Paris: Albatros, 1978).

the *dispositif* is understood as 'a thoroughly heterogeneous ensemble consisting of discourses, institutions, architectural forms, regulatory decisions, laws, administrative measures, scientific statements, philosophical, moral and philanthropic propositions—in short, the said as much as the unsaid'.[65] Furthermore, it includes coordination between human and non-human actors,[66] which is of particular interest in this book, as it incorporates both types of actors from the very beginning. By embracing this notion, it is therefore possible to address the vast system operating behind commercial radio stations, which are more than simple radio broadcasters. Moreover, as mentioned by Agamben, the *dispositif* always has a 'strategic function',[67] which, in the case of Radio Luxembourg and Europe n°1 could be found in their commercial nature.

The concept was later used in various disciplines, notably in communication studies, but, more importantly here, in media history. The concept was used, for instance, by Stefan Krebs[68] and in Tim van der Heijden's *Hybrid Histories*,[69] in regard to practices of listening in automobiles and for home movies respectively. Contemporary scholarship indeed made use of it as a conceptual tool to investigate various media; and, for instance, as a way to conceptualise technologies within the field of journalism,[70] while *Materializing Memories: Dispositifs, Generations, Amateurs*[71] explores the most relevant and recent works. As shown by the few examples discussed above, the *dispositif* is the ideal conceptual tool to encompass the various

[65] "The Confession of the Flesh" (1977), interview with Michel Foucault, in M. Foucault & C. Gordon (eds.), *Power/Knowledge: Selected Interviews and Other Writings* (Harlow: Longman, 1980), pp. 194–228.

[66] Larroche, *The Dispositif*, p. xv.

[67] Giorgio Agamben, "Théorie des dispositifs", *Po&sie* (115:1, 2006), pp. 25–33. Translation from Italian to French by Martin Rueff.

[68] Stefan Krebs, "'Sobbing, Whining, Rumbling': Listening to Automobiles as Social Practice", in T. Pinch & K. Bijsterveld (eds.), *The Oxford Handbook of Sound Studies*, pp. 79–101.

[69] Tim van der Heijden, *Hybrid Histories: Technologies of Memory and the Cultural Dynamics of Home Movies, 1895-2005* (Maastricht: Universitaire Pers Maastricht, 2018).

[70] Andreas Fickers, "'Neither Good, nor bad; nor neutral': The Historical *Dispositif* of Communication Technologies", in M. Schreiber & C. Zimmermann (eds.), *Journalism and Technological Change: Historical Perspectives, Contemporary Trends* (Frankfurt: Campus, 2014), pp. 30–52.

[71] S. Aasman, A. Fickers & J. Wachelder (eds.), *Materializing Memories: Dispositifs, Generations, Amateurs* (New York: Bloomsbury, 2018), pp. 1–15.

angles and perspectives chosen to study commercial radio stations in the Long 1960s.

What Are the Long Sixties? The focus of this book is on the so-called Long Sixties, the period covering the late 1950s to the early 1970s, which has been under the scrutiny of many historians and is still the focus of several works. Michael J. Heale even rightly suggests that 'why the era continues to excite such interest deserves a study of its own.'[72] As true as this statement is, this study is not on the agenda here, instead, the ways that this period has been framed in historical research are discussed. This magnetic attraction to the Sixties has led, as one could expect, to a rather important historiographical debate around the period, something that should be pondered upon, as 'the Sixties' can cover a wide range of definitions. The period is relevant to the history of commercial radio stations due, among other reasons, to its close links with the generational question of the 'baby boomers'. In his attempt to 'make sense of the Sixties',[73] Mark Lytle quotes Todd Gitlin, who explained that, in regards to the political events of the Sixties, 'there had to be a critical mass of students'.[74] This 'mass' is a reference to the baby- boomer generation, who also played a crucial role in the audience of commercial radio stations in the Sixties. Therefore, it means that strong ties between this specific generation and the period studied exist and should be kept in mind.

Numerous definitions of the timeframe applied to the 'Sixties' exist in historical literature. It appears that most historians reflecting on it start by expanding on the artificial nature of the decade and that the 'Sixties' are not, *stricto* sensu, the 1960s.[75] Authors sometimes stretch far from the strict decade, making the 'Sixties' go all the way from 1955 to 1975, as Elizabeth Martinez suggests, although the opposite vision, a much shorter one, also exists. The journalist Jon Margolis, for example, makes 'his'

[72] Michael J. Heale, "The Sixties as History: A Review of the Political Historiography", *Reviews in American History* (33:1, 2005), pp. 133–152.

[73] Mark Lytle, "Making Sense of the Sixties", *Irish Journal of American Studies* (10, 2001), pp. 1–17.

[74] Todd Gitlin, *The Sixties: Years of Hope, Days of Rage* (New York: Bantam Books, 1987).

[75] One example can be found in Arthur Marwick's introduction in Anthony Aldgate, James Chapman & Arthur Marwick (eds.), *Windows on the Sixties: Exploring Key Texts of Media and Culture* (London & New York: I. B. Tauris, 2000), p. xii.

Sixties start only in 1964,[76] while the historian Bruce Schulman, who wrote a book on the 1970s, thinks, on the other hand, that the Sixties ended in 1968—'a year of miracle and a year of horrors'[77]—to give space to the Seventies; or, at least, to his Seventies. These different examples of the Sixties reveal the fluidity of these limits, depending on the lens chosen to study the period. Following this idea, the most convincing timeframe suggested of 'the Sixties'—in the context of the history of commercial radio stations in Western Europe—is the one used by British historian Arthur Marwick, in his book on the topic,[78] from *circa* 1958 to *circa* 1974. It was chosen to apply a similar definition of the period throughout this book, as the scope chosen by Marwick is still relevant two decades later. His interest in a cultural approach, that covers multiple European countries and the United States, is particularly relevant in the context of this book.

The dates of 1964 and 1968 were used as demarcations in other works, as they represent key changes within the decade. Edward P. Morgan, already in 1991,[79] and, later on, Arthur Marwick, refer to the period 1964 to 1968 as the 'High Sixties'[80] Most of the debates about the framing of the 'Sixties' revolve around political questions. There is no doubt that 1968 was crucial for commercial radio stations, especially in regards to France,[81] however, these political moments are only so useful when it comes to a cultural approach to radio history. This being said, the 'High Sixties' is still a relevant concept for this book. 1966 was a year of crucial changes for the French service of Radio Luxembourg, which was renamed RTL. It also got new leadership, who launched a series of profound changes to the programmes and the studios. In Britain, the year

[76] Jon Margolis, *The Last Innocent Year: America in 1964* (New York: William Morrow and co., 1999).

[77] Bruce J. Schulman, *The Seventies: The Great Shift in American Culture, Society and Politics* (New York: Free Press, 2001), p. 1.

[78] Arthur Marwick, *The Sixties: Cultural Revolution in Britain, France, Italy and the United States, c.1958-c.1974* (Oxford, New York: Oxford University Press, 1998).

[79] Edward P. Morgan, *The 60s Experience: Hard Lessons About Modern America* (Philadelphia: Temple University Press, 1991).

[80] Marwick, *The Sixties*, p. 247.

[81] Richard Legay, "RTL & Europe n°1 as central actors. The importance of mobility for commercial radio stations during the Parisian events of Mai 68", *Rundfunk & Geschichte* (3:4, 2018), pp. 41–50.

1967 is linked to the Marine Offences Act that put a stop to the activities of prominent offshore radio stations, such as Radio Caroline, which had started its broadcast in 1964. BBC Radio One was created in 1967, with more light entertaining programmes and record shows, and both events had a strong impact on the media landscape, especially in Britain. These examples reveal that it makes sense to include the High Sixties in this book as they do represent a turning point, even from a more cultural approach. Moreover, it also allows to structure the reflection into three times, the Early Sixties (c. 1958 to 1964), the High Sixties (1964 to 1968), and the Late Sixties (1968 to c. 1974), which provides this work with a chronological structure that reveals itself useful.

Sources Consulted. This book relies on various sources, which can be divided into three main types developed in the following pages. First, institutional archives, such as minutes of a meeting and correspondence. Second, magazines published with the stations, in this case, *Fabulous 208* and *Salut les Copains*. Third, audio documents, more precisely recordings of some of the stations' shows, made by the institutions themselves or by amateurs. The use of such varied sources is what makes this book stand out. While the archives of RTL Group have been consulted by several historians, other archives—such as the maps found at the Felsberg transmitter site and the sound archives of Europe 1—have not. This has been particularly useful to shed new light on the history of commercial radio stations.

The analysis of commercial radio stations history includes, ideally, the stations' own archives, however, this endeavour can be particularly challenging.[82] In researching for this book, it was only possible to gain access, at RTL Group, to their 'paper' archives in Luxembourg, and, at Europe 1, only to their audio archives in Paris. It is important to take into account that these archives are not easily accessible due to the private and commercial nature of the institutions they belong to. Archival access has mostly been possible due to the goodwill of certain archivists, to whom the author is grateful. In order to supplement this inquiry, other institutions provided useful source material. Among others, the BBC Written Archives Centre in England, as well as the archives of a regulatory body

[82] Richard Legay, 'Studying 1960s commercial radio', in K. McDonald & H. Chignell (eds.), *Bloomsbury Handbook of Radio* (London: Bloomsbury, 2023), pp. 367–382.

such as the Independent Broadcasting Authority.[83] Several documents were consulted in the informal archives at the former transmitter of Europe n°1, located in the Saarland in Germany. A large number of maps, produced by the technical staff of the station, ranging from the 1950s to the 1970s, were found and digitised at the University of Luxembourg, allowing their analysis.

The research undertaken in this book also included a corpus of radio magazines published in collaboration with commercial radio stations, in this case, *Fabulous 208*[84] and *Salut les Copains*.[85] Complete, but non-digitised, collections of these magazine issues are available at the British Library and at the Bibliothèque nationale de France. Additionally, a selection of magazines was acquired[86] and digitised to facilitate their analysis, which relied on a textual analysis of articles published in the magazines, as well as other elements, such as letters and editorials. Due to the nature of the sources, images, in particular photos and photo-reportages, were analysed jointly with these texts.

A second corpus of sources for this book consisted of recordings of programmes that were broadcasted throughout the Sixties by the three chosen broadcasters: Radio Luxembourg/RTL (the French service), Radio Luxembourg 208 (the English service) and Europe n°1. Part of the material consulted was from archival institutions, such as the *Institut National de l'Audiovisuel* (known as INA) and the *Service de Documentation* of Europe 1, and some were gathered from recordings that were in the hands of amateurs and private collectors, such as former Europe n°1 technician Roland Biesen. For each broadcaster, around 20 to 30 hours worth of audio material were consulted.[87]

This book ponders upon the central hypothesis that commercial radio stations, specifically Radio Luxembourg and Europe n°1, throughout

[83] Available through the resources of Bournemouth University Library.

[84] More precisely, *Fabulous* was first published in 1964, and was renamed *Fabulous 208* after joining forces with Radio Luxembourg in 1966. As the focus is mostly on issues published after 1966, the name *Fabulous 208* is used throughout the book.

[85] It should be mentioned that commercial stations were involved in other types of publications, which are not covered in this book. Radio Luxembourg, for instance, published annual almanacks.

[86] By the author and by Aline Maldener from the University of Saarland, whose help was precious.

[87] See the appendix for more detail about the list of audio recordings consulted.

the long 1960s, have shaped and were shaped by a powerful *dispositif*. Defining what made this *dispositif*, and how it differentiated itself from public ones, is one of the core points of this work, which brings together radio programmes, sound material, institutional archives, and magazines. The research undertaken for this book was driven by this central query. Doing so, however, opened a series of other, more focused, questions which added to the general inquiry. First, how have Radio Luxembourg and Europe n°1, as transnational actors, appropriated, understood, and represented this broadcasting space? What was their place in it and what was their place in relation to other broadcasting actors? Second, what constituted the sound of commercial radio stations? What were they like, and how did this sound differ from other broadcasters? Third, what were the intermedial relations between radio stations and magazines? How did the two different media work together and support each other? Last, how were the three aspects of space, sound, and intermediality part of the *dispositif* of commercial radio stations and how did they contribute to it? These secondary questions are essential and are echoed in the way this book is structured.

Structure of the Book. This book is structured around 3 thematic parts. The first one is focused on the question of **space** and includes Chapters 2, 3, and 4. The transnational broadcasting space of Europe n°1 and Radio Luxembourg is at the heart of Chapter 2. It analyses where the space begins, at the transmitter sites, and ends, where the stations can no longer be heard; and how all of this changed throughout the Long Sixties. A study of the margins and the attempts to extend this space by including new markets is included. Chapter 3 is an analysis of the broadcasting space which is split into three intertwined layers—material, commercial, and imagined—which all offer a different perspective on the ways the stations perceived the space they were broadcasting to. In Chapter 4, several tensions within the transnational broadcasting space are analysed. First, it looks at the ways commercial radio stations played with a local focus in some of their programmes as well as displaying an international image. Then, it analyses the relations with other actors, such as offshore stations, the BBC, and the United States. Last, it looks at the competitive dynamic at stake between Radio Luxembourg and Europe n°1.

The second part, from Chapter 5 up to Chapter 7, is dedicated to the matter of **sound**. A detailed analysis of the sound of commercial radio in the Long Sixties is at the core of Chapter 5. It looks at nine recordings of light entertainment programmes, three per broadcaster, and three per sub-period of the Long Sixties. By doing so, this chapter highlights the similarity and the variations of the stations' programmes over time. Chapter 6 analyses a series of key features that define the sound of commercial radio stations. It is based on the analysis of a large corpus of audio recordings that shines light on elements that bind the sound of the stations together across languages and time. These features are commercialism, irreverence and playfulness, interactivity, liveness, and sonic icons. The chapter also includes a study of the ways the sound gives insights into the materiality of the radio experience. In Chapter 7, the unique joint programme, in French, between Radio Luxembourg and Europe n°1 in May 1968 is analysed as a case study. This humoristic stereophonic show of light entertainment was broadcast jointly by both stations and its study sheds light on aspects of cooperation between the stations, the ways they perceived themselves and each other, and how they represented, with humour, the media landscape in France at the time.

The third part, ranging from Chapter 8 to Chapter 10, deals with **intermediality**. First, Chapter 8 focuses on the intermedial links between commercial radio stations and magazines. More precisely, it looks at the connections between the English service of Radio Luxembourg and *Fabulous 208*, and between the magazine *Salut les Copains* and the eponym show on Europe n°1. After a brief history of the ways the links came to happen, this chapter shows how some cross-media features were shared between radio and magazines. Then, the importance of visual culture and stardom, omnipresent in the magazines, is studied in Chapter 9, which reveals how it impacted the overall *dispositif* of commercial radio. The role of photographs in the magazine, the importance of the creative consumer, and a case study of a small booklet called the *A-Z of Fab Fancies* are also addressed. Finally, in Chapter 10, the short-lived radio programme *Le Feu de camp du dimanche matin* is analysed. Broadcast on Europe n°1 in 1969, it was hosted by well-known names in the French-speaking comics industry. The analysis of this show makes a perfect case study of the intermedial connections between comics and radio.

A penultimate chapter is dedicated to an overarching analysis of the *dispositif* that binds commercial radio stations in the Long Sixties together. It details the ways the concept can be applied to Radio Luxembourg and Europe n°1 based on analyses conducted in the previous chapters. Made out of seven constitutive elements (inc. the studios and the transmitter sites), this chapter brings together the various dimensions, perspectives, and examples used throughout the book into an encompassing analysis. Last, Chapter 12 sets out the book's conclusions.

References

Aasman, Susan, Andreas Fickers and Joseph Wachelder (eds.). *Materializing Memories: Dispositifs, Generations, Amateurs*. New York: Bloomsbury, 2018.

Agamben, Giorgio. "Théorie des dispositifs", trans. Martin Rueff. *Po&sie* (115:1, 2006): 25–33.

Aldgate, Anthony, James Chapman and Arthur Marwick (eds.). *Windows on the Sixties: Exploring Key Texts of Media and Culture*. London & New York: I. B. Tauris, 2000.

Antoine, Frédéric. *Analyser la radio. Méthodes et mises en pratique*. Louvain-la-Neuve: De Boeck, 2016.

Badenoch, Alexander, Andreas Fickers and Christian Henrich-Franke (eds.). *Airy Curtains in the European Ether: Broadcasting and the Cold War*. Baden-Baden: Nomos, 2013.

Baudry, Jean-Louis. *L'Effet cinéma*. Paris: Albatros, 1978.

Beccarelli, Marine. *Micros de nuit: histoire de la radio nocturne en France, 1945-2012*. Rennes: Presses Universitaires de Rennes, 2021.

Beccarelli, Marine. *La Nuit du bout des ondes*. Paris: INA Editions, 2014.Gitlin, Todd. *The Sixties: Years of Hope, Days of Rage*. New York: Bantam Books, 1987.

Berg, Katja. *Grenzenlose Unterhaltung: Radio Luxemburg in der Bundesrepublik, 1957-1980*. Göttingen: Wallstein Verlag, 2021.

Berg, Katja and Anna Jehle. "'Through the Air to Anywhere': Radio Luxembourg—A Transnational Broadcasting Station?" In *Transnational Broadcasting in Europe 1945–1990*, edited by Christoph Classen, 23–44. Frankfurt am Main: Peter Lang, 2016.

Bernard, Luc. *Europe 1: La Grande histoire dans une grande radio*. Paris: Centurion, 1990.

Birdsall, Carolyn. *Nazi Soundscapes: Sound, Technology and Urban Spaces in Germany, 1933–1945*. Amsterdam: Amsterdam University Press, 2012.

Blandin, Claire. "Radio et magazine: une offre plurimédia pour les jeunes des Sixties". *Le Temps des Médias* 21:2 (2013): 134–142.

Bull, Michael. *Sounding Out the City*. 2nd ed. Oxford: Berg, 2006.

Cheval, Jean-Jacques. *Les Radios en France: Histoire, état et enjeux*. Rennes: Apogée, 1997.

Chignell, Hugh. *Key Concepts in Radio Studies*. London: Sage, 2009.

Classen, Christoph (ed.). *Transnational Broadcasting in Europe 1945-1990*. Frankfurt am Main: Peter Lang, 2016.

Corbin, Alain. *Les Cloches de la terre: paysage sonore et culture sensible dans les campagnes au XIXe siècle*. Paris: Flammarion, 1994.

Corbin, Alain. *Le Miasme et la jonquille: l'odorat et l'imaginaire social, XVIIIe-XIXe siècles*. Paris: Flammarion, 1982.

Cronqvist, Marie and Christoph Hilgert. "Entangled Media Histories. The Value of Transnational and Transmedial Approaches in Media Historiography". *Media History* 23:1 (2017): 130–141.

Cronqvist, Marie and Christoph Hilgert. "Entangled Media Histories: Response to the Responses". *Media History* 23:1 (2017): 148–149.

Defrain, Jean-Pierre and Jacques Boutelet. *RTL. 40 ans ensemble*. Paris: Calmann-Lévy, 2006.

Dominguez Muller, David. *Radio-Luxembourg: Histoire d'un média privé d'envergure européenne*. Paris: L'Harmattan, 2007.

Douglas, Susan J. *Listening In: Radio and the American Imagination*. New York: Times Books, 1999.

Febvre, Lucien. *A New Kind of History: From the Writings of Febvre*. Translation by K. Folca. New York: Harper and Row, 1973.

Fesneau, Elvina. *Le Transistor à la conquête de la France. La radio nomade (1954–1970)*. Paris: INA Editions, 2014.

Fickers, Andreas and Pascal Griset. *Communicating Europe: Technologies, Information, Events*. Basingstoke: Palgrave Macmillan, 2019.

Fickers, Andreas. "'Neither Good, nor bad; nor neutral': The Historical *Dispositif* of Communication Technologies". In *Journalism and Technological Change: Historical Perspectives, Contemporary Trends*, edited by Martin Schreiber and Clemens Zimmermann, 30–52. Frankfurt: Campus, 2014.

Fickers, Andreas. "Broadcasting". In *Palgrave Dictionary of Transnational History*, edited by Akira Iriye and Jean-Yves Saunier, 106-108. Basingtoke: Palgrave Macmillan, 2009.

Fickers, Andreas. "Radio". In *Palgrave Dictionary of Transnational History*, edited by Akira Iriye and Jean-Yves Saunier, 870–873. Basingtoke: Palgrave Macmillan, 2009.

Fickers, Andreas. *Der 'Transistor' als technisches und kulturelles Phänomen: Die Transistorisierung der Radio- und Fernsehempfänger in der deutschen Rundfunkindustrie, 1955–1965*. Bassum: Verlag für Geschichte der Naturwissenschaften und Technik, 1998.

Foucault, Michel. "The Confession of the Flesh". In *Power/Knowledge: Selected Interviews and Other Writings*, edited by Michel Foucault and Colin Gordon, 194–228. Harlow: Longman, 1980.

Griffen-Foley, Bridget. "Entangled Media Histories: A Response". *Media History* 23:1 (2017): 145–147.

Heale, Michael J. "The Sixties as History: A Review of the Political Historiography". *Reviews in American History* 33:1 (2005): 133–152.

Heijden, Tim van der. *Hybrid Histories: Technologies of Memory and the Cultural Dynamics of Home Movies, 1895–2005*. Maastricht: Universitaire Pers Maastricht, 2018.

Hilgert, Christoph, Marie Cronqvist and Hugh Chignell. "Tracing Entanglements in Media History". *Media History* 26:1 (2020): 1–5.

Hilmes, Michele. "Entangled Media Histories: A Response". *Media History* 23:1 (2017): 142–144.

Hilmes, Michele. *Network Nations: A Transnational History of British and American Broadcasting*. London: Routledge, 2012.

Hilmes, Michele. *Radio Voices: American Broadcasting, 1922–1952*. Minneapolis: University of Minnesota, 1997.

Hochscherf, Tobias, Richard Legay and Hedwig Wagner. "Radio Beyond Boundaries: An Introduction", *Historical Journal of Film, Radio and Television* 39:3 (2019): 431–438.

Iriye, Akira. *Global and Transnational History*. Basingstoke: Palgrave Macmillan, 2013.

Iriye, Akira and Jean-Yves Saunier (eds.). *Palgrave Dictionary of Transnational History*. Basingtoke: Palgrave Macmillan, 2009.

Jehle, Anna. *Welle der Konsumgesellschaft: Radio Luxembourg in Frankreich, 1945–1975*. Göttingen: Wallstein Verlag, 2018.

Jenks, John. "Commercial temptation: Cross-border radio and the comparative transformation of public service broadcast policy in Britain, South Africa and India, 1930–67". *Radio Journal: International Studies in Broadcast & Audio Media* 15:1 (2017): 125–141.

Jouin, Gilbert, *Salut les Copains*. Paris: Ipanema, 2012.

Keith, Michael C. "The Long Road to Radio Studies". *Journal of Broadcasting & Electronic Media* 51:3 (2007): 530–536.

Kessler, Frank. "Notes on *Dispositif*". Online, 2007. [http://www.frankkessler.nl/wp-content/uploads/2010/05/Dispositif-Notes.pdf]

Krebs, Stefan. "'Sobbing, Whining, Rumbling': Listening to Automobiles as Social Practice". In *The Oxford Handbook of Sound Studies*, edited by Trevor Pinch and Karin Bijsterveld, 79–101. Oxford: Oxford University Press, 2012.

Lacey, Kate. "Up in the air? The matter of radio studies". *The Radio Journal: International Studies in Broadcast & Audio Media* 16:2 (2018): 109–126.

Lacey, Kate. "Ten Years of Radio Studies: The Very Idea". *The Radio Journal: International Studies in Broadcast & Audio Media* 6:1 (2008): 21–32.

Langenbruch, Anna (ed.). *Klang als Geschichtsmedium: Perspektiven für eine auditive Geschichtsschreibung*. Bielefeld: Transcript, 2018.

Larroche, Valérie. *The Dispositif: A Concept for Information and Communication Sciences*. London: ISTE, 2019.

Legay, Richard. "Studying 1960s commercial radio". In *Bloomsbury Handbook of Radio*, edited by McDonald, Kathryn and Hugh Chignell, 367–382. London: Bloomsbury, 2023.

Legay, Richard. "RTL & Europe n°1 as central actors. The importance of mobility for commercial radio stations during the Parisian events of Mai 68". *Rundfunk & Geschichte* 3:4 (2018): 41–50.

Lommers, Suzanne. *Europe—On Air. Interwar Projects for Radio Broadcasting*. Amsterdam: Amsterdam University Press, 2012.

Loriou, Céline. "Faire de l'histoire, un casque sur les oreilles: le goût de l'archive radiophonique". In *Le Goût de l'archive à l'ère numérique*, edited by Frédéric Clavert and Caroline Muller. Online, 2018. [http://www.gout-numerique.net/table-of-contents/faire-de-lhistoire-un-casque-sur-les-oreilles-le-gout-de-larchive-radiophonique]

Lytle, Mark, "Making Sense of the Sixties". *Irish Journal of American Studies* 10 (2001): 1–17.

Maréchal, Denis. *RTL. Histoire d'une radio populaire: De Radio Luxembourg à RTL.fr*. Paris: Nouveau Monde Éditions, 2010.

Maréchal, Denis. *Europe 1: de la singularité au déclin (1955–2022)*. Lormont: Bord de l'Eau, 2023.

Maréchal, Denis. *Radio-Luxembourg 1933–1993, un média au coeur de l'Europe*. Nancy: Presses Universitaires de Nancy, 1994.

Margolis, Jon. *The Last Innocent Year: America in 1964*. New York: William Morrow and co., 1999.

Marwick, Arthur. *The Sixties: Cultural Revolution in Britain, France, Italy and the United States, c.1958-c.1974*. Oxford, New York: Oxford University Press, 1998.

Newton, Gerald. "Radio Luxembourg in Peace and War". *German Life and Letters* 66:1 (2013): 55–76.

Nicholas, Sîan. "Media History or Media Histories? Re-addressing the History of Mass-media in Inter-war Britain". *Media History* 18:3-4 (2012): 379–394.

Nichols, Richard. *Radio Luxembourg: The Station of the Stars*. London: W.H. Allen, 1983.

Olivennes, Denis, Franck Ferrand and Bruno Labous. *Europe 1: Le Dictionnaire amoureux illustré*. Paris: Plon, 2015.

Pinch, Trevor and Karin Bijsterveld (eds.). *The Oxford Handbook of Sound Studies*. Oxford: Oxford University Press, 2012.

Quillien, Christophe. *Nos Années Salut les Copains 1959–1976* (Paris: Flammarion, 2009)

Savage, Barbara D. *Broadcasting Freedom*. Chapel Hill: The University of North Carolina Press, 1999.

Schafer, R. Murray. *The Soundscape: Our Sonic Environment and the Tuning of the World*, 2nd ed. New York: Rochester, 1994.

Schulman, Bruce J. *The Seventies: The Great Shift in American Culture, Society and Politics*. New York: Free Press, 2001.

Spohrer, Jennifer. "Ruling the Airwaves: Radio Luxembourg and the Origins of European National Broadcasting, 1929–1950". PhD thesis., Columbia University, 2008.

Sterne, Jonathan (ed.). *The Sound Studies Reader*. Abingdon-on-Thames: Routledge, 2012.

Sterling, Christopher H. and John Michael Kittross. *Stay Tuned: A History of American Broadcasting*, 3rd ed. Mahwah & London: Lawrence Erlbaum Associates, 2002.

Street, Seán. *Crossing the Ether: Pre-War Public Service Radio and Commercial Competition, 1922–1945*. Eastleigh: John Libbey Publishing, 2006.

Suisman, David. "Introduction: Thinking Historically about Sound and Sense". In *Sound in the Age of Mechanical Reproduction*, edited by David Suisman and Susan Strasser, 1–12. Philadelphia: University of Pennsylvania Press, 2010.

Tinker, Chris. "Mixed Masculinities in 1960s British and French Youth Magazines". *The Journal of Popular Culture* 7:1 (2014): 84–108.

Tinker, Chris. "Shaping 1960s youth in Britain and France: Fabulous and Salut les Copains". *International Journal of Cultural Studies* 14:6 (2011): 641–657.

Tinker, Chris. *Mixed Messages: Youth Magazines Discourse and Sociocultural Shifts in Salut les Copains (1962–1976)*. Oxford, New York: Peter Lang, 2010.

Tinker, Chris. "Shaping Youth in Salut les Copains". *Modern & Contemporary France* 15:3 (2007): 293–308.

Wagner, Hans-Ulrich, Hugh Chignell, Marie Cronqvist, Christoph Hilgert and Kristin Skoog. "Media after 1945: Continuities and New Beginnings". In *The Handbook of European Communication History*, edited by Klaus Arnold, Paschal Preston and Susanne Kinnebrock, 189–204. London: Wiley, 2020.

PART I

Space

Radio Luxembourg and Europe n°1 are quintessentially transnational actors. Their nature—as broadcasters—is already, in essence, a transnational one, as radio's 'technical opportunity to operate beyond the confines of individual countries' denotes.[1] Moreover, the thrill of listening to faraway stations was fascinating for early radio enthusiasts,[2] who engaged in a transcendence of 'geographical space by communing not with those next to them but with the "distant" voices transmitted through the ether',[3] also inferring this transnational nature from the user's perspective. In the case of these commercial stations, however, transnationalism is not a by-product of their nature, it is their main *raison d'être*. Both Radio Luxembourg, and, later, Europe n°1, were born of the specific geopolitical situation and broadcasting regulations of their time, which led to their installation in the Grand Duchy of Luxembourg and in the Saarland. From there, they could broadcast to, and benefit from, the large markets of neighbouring countries, where their transmitters were

[1] T. Hochscherf, R. Legay & H. Wagner, "Radio Beyond Boundaries: An Introduction", *Historical Journal of Film, Radio & Television* (39-3, 2019), pp. 431–438.

[2] T. D. Taylor, M. Katz & T. Grajeda (eds.), *Music, Sound, and Technology in America: A Documentary History of Early Phonograph, Cinema, and Radio* (Durham: Duke University Press, 2012), p. 240.

[3] Veit Erlmann, "But What of the Ethnographic Ear?", in Veit Erlmann (ed.), *Hearing Cultures. Essays on Sound, Listening and Modernity* (Oxford: Berg, 2004), pp. 1–20.

not allowed at the time. National borders are not overlooked by a transnational lens, they even play a crucial role in the apparition and existence of commercial stations, which used them to circumvent national regulations and monopolies on broadcasting. It appears that transnationalism plays a prominent role for commercial stations, and, therefore, deserves to be a study focus in this book. This first part, made of three chapters, thus investigates the relationship between commercial stations and the space in which they operated. Based on these observations on their nature and on their origins, it appears that transnationalism goes hand in hand with commercial stations, and, interestingly, this transnational influence seems to have manifested itself even in the smallest details of everyday objects.

It can be argued that the presence of an imagined transnational radio space is already suggested in the very device used to listen to it. Indeed, the radio dial, present on receivers throughout the Sixties, is part of the appropriation, by the listener, of an imagined European broadcast landscape.[4] The commercial radio stations studied in this book were no exception, as many receivers bore a mention to Luxembourg, and to Europe 1. The stations were therefore stops in the 'imagined voyage through the ether' suggested by the dial, an element that was part of the domestication of radio as a mass medium[5] because it made the interface of the receiver more user-friendly. Through the names on the dial, many European countries and cities were present and their transnational airwaves were brought to listeners. Moreover, as Fickers' chapter shows, a regulatory institution, the International Broadcasting Union,[6] was responsible for the implementation of calibrated station scales, therefore homogenising the process at a large scale, reinforcing the transnational suggestive power of radio dials.

[4] Andreas Fickers, "Visibly Audible. The Radio Dial as Mediating Interface", in Trevor Pinch & Karin Bijsterveld (eds.), *Oxford Handbook of Sound Studies* (Oxford: Oxford University Press, 2012), pp. 411–439.

[5] Two monographs for France and the United States are in this regard helpful. See Cécile Méadel, *Histoire de la radio des années trente : du sans-filiste à l'auditeur* (Paris: Anthropos, 1994) & Michele Hilmes, *Radio Voices: American Broadcasting, 1922–1952* (Minneapolis: University of Minnesota Press, 1997).

[6] The International Broadcasting Union, known as the IBU, was an institution gathering European radio broadcasters. Established in 1925, its goal was to resolve international issues related to broadcasting. It was dissolved in 1950 and subsequently replaced by the EBU, the European Broadcasting Union.

The power of radio dials to conjure the transnational imagination of the user-listener was not limited to visual and material elements: the sound was also intertwined in this suggestive potential, through the 'complex interplay of the haptic and the hearing sense'.[7] Such a sonic illustration of the transnational potential of radio is present in a handful of audio recordings found at the INAthèque in Paris. In those sources from 1962 and 1963, an unknown listener recorded evening news, in French, of Radio Luxembourg and Europe n°1, switching from one station to the other. In one of the sources from 1963,[8] one can hear Europe n°1's iconic jingle—its key sonic icon—with trumpets and other brass instruments. It is followed by a news report, and lengthy details about the Bastille Day celebrations earlier in the day, which include interviews of passersby, and are interspersed with adverts for razors and peaches. Suddenly, voices stop, and after a short silence, a different journalist reads a news report, mentioning his colleague Christian Brincourt,[9] identifying the station as Radio Luxembourg. The news content is quite similar to those of the previous segment, and other adverts can be heard, however, the sound quality of both stations is clearly different. Radio Luxembourg can be heard in a far clearer way than Europe n°1, whose sound quality is, to put plainly, of a poorer quality. The volume of the recording fluctuates, a voice from another station can be heard faintly in the background, and occasional interferences are present. These audio elements, as well as the sound of changing frequencies—short silence included[10]—between the two stations constitute fascinating illustrations of the transnational

[7] Andreas Fickers, "The Radio Station Scale: A Materialized European Event", in A. Badenoch & A. Fickers (eds.), *Materializing Europe: Transnational infrastructures and the Project of Europe* (Basingstoke: Palgrave Macmillan, 2014), pp. 252–254.

[8] Journaux parlés de Radio Luxembourg et d'Europe n°1, Radio Luxembourg & Europe n°1, 14th July 1963, 00:23:12, Inathèque, Paris.

[9] Christian Brincourt, born in 1935, was a French radio reporter, who joined Radio Luxembourg in 1960. He notably covered the war in Algeria, the Six-Day war and the riots of May 1968 in France.

[10] Silence is not the absence of sound, it is a highly significant element, and therefore crucial to indicate change. See David Le Breton, *Du Silence* (Paris: Éditions Métailié, 2015), p. 152.

suggestion of radio.[11] In this recording, the act of changing stations—therefore, changing countries, from Germany to Luxembourg—is not only suggested by the radio dial but heard. The two transnational stations clearly sound different, suggesting technical differences in the European broadcast landscape.

Radio dials were more than an interface between listeners and stations. They also played a crucial role for broadcasters for broadcasters and for manufacturers,[12] who used them to promote their receivers. Directors of Radio Luxembourg appeared glad, in 1971, to learn that the power boost of their short-wave transmitter led to a clear amelioration of the listening quality of their German programmes, which pushed manufacturing companies to implement buttons bearing the name of Radio Luxembourg on their radio sets.[13] Such buttons on the dials facilitated the reception of the commercial station, and clearly identified it within the radio landscape. Moreover, with 22 million radio sets in the Federal Republic able to pick up short waves in 1969,[14] Radio Luxembourg hoped to see an increase in its listenership thanks to this new button. This high level of transistor ownership was also found in France, where more than 70% of households owned at least one in 1969.[15]

Therefore, the radio dial, one small but integral part of every radio set, was a gateway to transnationalism for listeners. One can already find in the radio dial the three layers—technical, commercial, and imagined—that structure Chapter 2. The dial allowed for a better appropriation by the listener of radio waves. It also represented a commercial stake for

[11] As mentioned by Street: 'a very small sound, given knowledge and the context of its history, becomes hugely significant. And this is what occurs with the almost insignificant sounds of changing frequencies, which are particularly "loud" to broadcasting historians.' Sean Street, *Sound at the Edge of Perception: The Aural Minutiae of Sand and other Worldly Murmurings* (Basingstoke: Palgrave Macmillan, 2019), p. 4.

[12] Andreas Fickers, "Visibly Audible", in Pinch & Bijsterveld, *The Oxford Handbook of Sound Studies*, pp. 411–439.

[13] Minutes of the "Assemblée générale annuelle", 18th May 1971, RTL Group Archives, Luxembourg.

[14] Minutes of the "Comité de direction", 3rd October 1969, RTL Group Archives, Luxembourg.

[15] Elvina Fesneau, *Le Poste à transistors à la conquête de la France* (Bry-sur-Marne: INA Editions, 2011), p. 221.

broadcasters and was an entry point for this 'imagined voyage in the ether'.[16]

Commercial radio stations could be listened to in many countries, some far away from the Saarland and the Grand Duchy, all the way to Scandinavia and Eastern Europe, however, the following chapters focus on Western European countries, which were at the core of the stations' markets. Thus Britain, West Germany, Belgium, Luxembourg, and the Netherlands are the prominent territories considered. France, the biggest market and main region of the competition between Europe n°1 and Radio Luxembourg, is studied with even more scrutiny, as it is a country, particularly highlighting the complexity of the dynamics at play.

[16] Moreover, as shown by Moores and Bull, communication technologies have allowed listeners to 'travel' and picture themselves as members of national and transnational communities, while simultaneously symbolically shaping the construction of 'home'. See Shaun Moores, *Interpreting Audiences: The Ethnography of Media Consumption* (London: Sage, 1993), p. 22, & Michael Bull, "Sound, Proximity, and Distance in Western Experience", in Erlmann, *Hearing Cultures*, pp. 173–190.

CHAPTER 2

A Far-Reaching Transnational Space

Britain, West Germany, the countries of the Benelux and France were at the heart of the space constructed and appropriated by commercial radio stations; they were the key markets and countries towards which the transmitters were broadcasting. Therefore, they are, as mentioned previously, the key focus of the study undertaken in this book. That is why this chapter analyses the 'edges' of the transnational space, to help define them before the following chapter returns to the key markets. What is meant by 'edges' here are the two ends of the space: where radio, so to say, happened, and where it disappeared. More precisely, this chapter analyses the transnational dynamics at play within the heart of the space, by looking into the connections between the stations and their transmitters and studios. In other words, the material anchoring enabled Radio Luxembourg and Europe n°1 to broadcast to a large space. The broad nature of the studied commercial radio stations implies that this space was not limited to a handful of countries. Radio waves can cross national borders, and the entrepreneurial spirit of commercial stations pushed them to look for new markets and audiences, sometimes in other countries. These other 'edges'—these regions and groups at the ends of the studied space—are therefore considered in this chapter. They were margins, countries at the very end of the technical reach of the stations, as well as at the very end of commercial interests. There were also short-lived, or even aborted, regions and groups that the stations reached,

R. Legay, *The Cultural Histories of Radio Luxembourg and Europe n°1*,
https://doi.org/10.1007/978-3-031-46250-4_2

or envisaged to reach. This chapter clearly sets the boundaries of the transnational space, by operating an analysis of its framework.

1 The Material Anchoring of Commercial Radio Stations

At first glance, radio might appear as an immaterial medium. And, even if there is some truth to this statement, there is a strong tangible dimension to radio. Beyond sound and airwaves, the more physical elements, such as the places where radio was made, should not be omitted. These can refer to studios and transmitters, but also to offices, and it can be argued that these locations, and the ways they connect with one another, constituted the foundation of the transnational potential of commercial radio stations. As detailed below, Radio Luxembourg and Europe n°1 relied on having these locations in different countries to operate, therefore implementing transnational dynamics. Looking at Iriye and Saunier's definition of transnational history, presented as an interest in 'links and flows' and a track of 'people, ideas, products, processes and patterns that operate over, across, through, beyond, above, under, or in-between politics and societies',[1] the stations offer material which fits in well with this definition. The flows and dynamics, here, are understood as all communications, people and products exchanged, across borders, between the different locations; locations which are at the centre of the following pages.

Both commercial radio stations relied on powerful transmitters to broadcast their programmes to a large audience. In the case of Europe n°1, founded in 1955 by Charles Michelson and Louis Merlin, with offices in Paris, rue François Ier, the transmitter site was installed in the Saarland, on the Felsberg plateau, where antennas for longwave broadcast were built.[2] From this position, near the Franco-German border, it was possible to be heard throughout metropolitan France.[3] The Saarland

[1] Akira Iriye & Pierre-Yves Saunier (eds.), *The Palgrave Dictionary of Transnational History* (Basingstoke: Palgrave Macmillan, 2009), p. xviii.

[2] Europe n°1 broadcast on LW from 1955 until the 31st of December 2019, slightly before midnight. See Sophie Schülke, "Endgültiger Sendeschluss in Berus", *Saarbrücker Zeitung*, 17th January, 2020.

[3] The first broadcasts of the station were, however, controversial ones as they interfered with other broadcasters that led to international debates. See Andreas Fickers,

is a particularly interesting region for transnational history, at the crossroads between France and Germany, and it was, at the time, under French administration. The creation of Europe n°1 stemmed from the short-lived commercial station Tele-Saar, which was itself inspired by the success of Radio Luxembourg.[4]

Radio Luxembourg, founded in 1933, had its transmitters in the Grand Duchy of Luxembourg, more precisely in Marnach and in Junglinster. One particularity of the station was that its antennas broadcasted two different programmes.[5] One in French,[6] and one in various languages (English, Flemish, German,[7] and Luxembourgish). In the 1960s, the programmes in French had studios and offices in Paris, at 22, rue Bayard.[8] Programmes in other languages were mostly made in the studios at the Villa Louvigny,[9] in the city of Luxembourg, which was also the headquarters of the station, and of the company behind it. Radio Luxembourg London Ltd., responsible for the English programmes, was based in London, on Hertford Street. There, a few dozen people worked on advertising sales and preparing tapes that would then be broadcast

"Die Anfänge des kommerziellen Rundfunks im Saarland", in C. Zimmermann, R. Hudemann & M. Kuderna (eds.), *Medienlandschaft Saar von 1945 bis in die Gegenwart. Band 1: Medien zwischen Demokratisierung und Kontrolle (1945–1955)* (München: R. Oldenbourg, 2010), pp. 241–308.

[4] Andreas Fickers & Pascal Griset, *Communicating Europe. Technologies, Information, Events* (Basingtoke: Palgrave Macmillan, 2019), p. 214.

[5] In this regard, the role of Mathias Felten (1907–1992) should be mentioned. An engineer from Luxembourg, Felten was a key figure in the early days of Radio Luxembourg, and held several prominent positions in the *Compagnie Luxembourgeoise de Télédiffusion*, or CLT, the company behind Radio Luxembourg. Moreover, he was partly responsible for the introduction of German language programmes in 1957.

[6] See Anna Jehle, *Welle der Konsumgesellschaft. Radio Luxembourg in Frankreich, 1945–1975* (Göttingen: Wallstein, 2018).

[7] See Katja Berg, *Grenzenlose Unterhaltung. Radio Luxemburg in der Bundesrepublik, 1957–1980* (Göttingen: Wallstein, 2021).

[8] Documents found in the archives of RTL Group reveal how important it was for the station to be visible in Paris, and to maintain the same address on Rue Bayard, as it was "associated in listeners' mind to Radio Luxembourg". See Minutes of the "Comité de direction", 27th June 1966, RTL Group Archives, Luxembourg.

[9] With the notable exception of Georges Lang's night programme, *Les Nocturnes*, launched in 1973, and which was made at the Villa Louvigny but broadcast on the French service from 1973. Marine Beccarelli, "Micros de nuit: Histoire de la radio nocturne en France, 1945–2012", PhD thesis. (Université Paris-1 Panthéon Sorbonne, 2016), p. 169.

from the Villa Louvigny, where English-speaking hosts would complete the programme.[10] Therefore, by relying on networks of underground cables,[11] telephone lines and postal services, commercial radio stations had almost constant connections between their studios, offices, and transmitters. These strong transnational flows enabled them to broadcast to their different markets in metropolitan France, the United Kingdom and Ireland, the Federal Republic of Germany, Belgium, and the Netherlands.

This brief description of commercial stations' communication networks suggests the importance of communication technology in the establishment of these private powers. It also reveals the importance of a multi-level approach[12] to embracing how local dynamics can impact the reach of the stations to a larger transnational space, all the way to Eastern Europe and Scandinavia, from areas in the peripheries of larger markets.

2 Broadcasting Across the Iron Curtain: Czechoslovakia and Poland

The nature of radio broadcasting meant that waves not only crossed the borders of targeted countries but also reached beyond. In addition to this, these areas were often at the end of the technical reach of transmitters, therefore putting them in the position of margins. Some of these margins in commercial radio stations' broadcasting space were found on the other side of the Iron Curtain, in Eastern Europe. Historical literature has shown that connections existed between the two blocs, deconstructing the idea of two separated communication entities,[13] and the same thing occurs here, as commercial stations could be heard in the East. In the case of television, there was an asymmetrical relationship between the two

[10] Denis Maréchal, *RTL. Histoire d'une radio populaire* (Paris: Nouveau Monde, 2010), p. 183.

[11] Katja Berg & Anna Jehle, " 'Through the Air to Anywhere': Radio Luxembourg—A Transnational Broadcasting Station?", in C. Classen (ed.), *Transnational broadcasting in Europe, 1945-1990* (Frankfurt am Main: PL Academic Research, 2016), pp. 23–44.

[12] Fickers & Griset, *Communicating Europe*, p. 5.

[13] A. Badenoch, A. Fickers & C. Henrich-Franke, *Airy Curtains in the European Ether. Broadcasting and the Cold War* (Baden-baden: Nomos, 2013), p. 9.

spheres,[14] and, a similar dynamic seems to apply in the context of this chapter. Indeed, the stations had listeners in the East, and seemed aware of the fact, although they never considered turning to those margins, likely because there was no conceivable financial gain there.

In a book they wrote together, Tony Prince and Jan Seslak shared some memories and anecdotes about their relationship to radio and, more specifically, to the English service of Radio Luxembourg, from both sides of the Cold War. Prince was a disc-jockey on Radio Caroline and Radio Luxembourg, while Seslak was a radio enthusiast, who listened to Western stations from his homeland of Czechoslovakia. He recalls for instance tuning in to Voice of America and the 'rather incredible feeling to know that I was listening and understanding fully a programme in English, as were millions of other people around the world.'[15] His listening habits in the Late Sixties, however, turned towards Radio 208, in particular the shows hosted by Prince, whom he managed to contact. Following an invitation from Seslak, the British disc-jockey came to Czechoslovakia in July 1970 to play records in clubs in three different towns around the country.[16] The anecdotes shared in the book are particularly interesting as they illustrate how a Czechoslovak citizen developed a taste for Western radio stations, giving a more personal and individual flavour to a known historical phenomenon. The visit of the Prince to the Eastern Bloc points to a permeability of the Iron Curtain that is not simply for airwaves, but also, on this occasion, for a representative of a Western station.

A second example can be found in Poland, where many people tuned in to the English service of Luxembourg in the 1960s and 1970s. An exhibition was put together in 2011 between the Embassy of Luxembourg in Poland and the History department at the University of Warsaw, which focused on the memories of Radio Luxembourg in Poland during the socialist period. The exhibition was also turned into a short book, in

[14] Andreas Fickers, "Looking East-Watching West? On the Asymmetrical Interdependencies of Cold War European Communication Spaces", in K. Bönker, J. Obertreis & S. Grampp (eds.), *Television Beyond and Across the Iron Curtain* (Newcastle upon Tyne: Cambridge Scholars Publishing, 2016), pp. 2–24.

[15] Tony Prince & Jan Seslak, *The Royal Ruler and the Railway DJ* (Slough: DMC Publishing Ltd., 2016), p. 87.

[16] ibid., p. 426.

English and in Polish offering fascinating insights into the work accomplished to put together the exhibition.[17] According to this book, Radio Luxembourg appears to have been tolerated by the socialist government—meaning the signal was not jammed—as the station was not perceived as political, as opposed to Voice of America and Radio Free Europe for instance. Many testimonies published in the book and in the exhibition discuss the act of listening, tuning in, and struggling to find the right frequency and right spot to have the best quality, due, among other things, to the distance between Poland and the transmitter in Luxembourg. This difficulty in listening to the station, in addition to the recurrent image in the testimonies of "opening a window to the West",[18] contribute to this idea of a far-reaching transnational space, Poland being in the margins of Radio Luxembourg's audience, in a geographical, technical, and commercial sense.

The two examples mentioned above are illustrations of the permeability of the Iron Curtain, in regards to broadcasts from the West. Many countries and societies in the Eastern bloc were indeed able to listen to stations from the West. The study of this large transnational broadcasting phenomenon led to the idea of an 'Airy Curtain' to illustrate this permeability, as well as the division of communication spaces in Europe, during the Cold War, which did not match material borders of the time.[19] The United States funded stations broadcasting towards Eastern countries, such as Voice of America, Radio Free Europe, and Radio Liberty. An effort supported, to a lesser extent, by the United Kingdom, France, and the Federal Republic of Germany.[20] In response, the Soviet Union tried to close its cultural borders and limit the impact of these stations, which were seen as foreign subversive agents of 'psychological warfare',[21]

[17] Milena Abdel-Massih & Anna Malewska-Szałygin, *Remembering Radio Luxembourg in the People's Republic of Poland* (Warsaw: Coma2, 2012).

[18] Other similar quotes include: "Radio Luxembourg was a window onto the world"; "This radio station was a kind of musical window onto the world"; "one of the few bonds which connected us to the outside world"; "Over there was a different world". Abdel-Massih & Malewska-Szałygin, *Remembering Radio Luxembourg*, pp. 20–21.

[19] Badenoch, Fickers & Henrich-Franke, *Airy Curtains*, p. 16.

[20] Tristan Mattelart, *Le Cheval de Troie audiovisuel: Le Rideau de Fer à l'épreuve des Radios et Télévisions Transfrontières* (Grenoble: Presses Universitaires de Grenoble, 1995), p. 42.

[21] ibid., pp. 29–30.

in which entertainment played a crucial role. The English service of Radio Luxembourg, at the heart of the examples developed above, is also mentioned in Mattelart's monograph as a broadcaster that provided Eastern European youth with recent hits and introduced them to private radio formats. However, 'if the entertainment programmes of the station had a similar impact in terms of propaganda than those of Radio Free Europe or Voice of America, they are not used for this purpose. This propaganda function results simply from the overflowing of borders [...]. The role of Radio Luxembourg as a means for entertainment in Eastern Europe is nevertheless non-negligible'.[22] This statement shows the importance of transnational broadcasting, sometimes even above the content broadcasted.

3 Margins and Short-Lived Attempts

In addition to the Eastern European margins, commercial stations made a few short-lived transnational attempts—with more or less success—to reach different countries and audiences, by broadcasting in other languages and to foreign places. A series of examples from Radio Luxembourg ire developed as illustrations of this phenomenon and to reflect its diversity.

The English service of Radio Luxembourg—or 208—was known to be listened to in Scandinavia throughout the period. One particular occasion highlighted this listening habit; the broadcasting of a boxing match in 1959, which was strongly followed in 208 by Swedish listeners.[23] This event strongly stands out as the station does not appear to have broadcasted specifically for a Swedish audience on other occasions, however, it sheds light on the non-negligible presence of listeners in Scandinavia. In 1951, out of the 13 million listeners of the English programme on Radio Luxembourg, 5 million were from Scandinavia, according to numbers

[22] ibid., p. 150.

[23] The boxing match opposed the American Floyd Patterson (1935–2006) to the Swedish Ingemar Johansson (1932–2009) on 26th June 1959, at the Yankee Stadium in New-York. Johansson defeated Patterson, therefore winning the prestigious world heavyweight champion title. The Swedish state-radio did not broadcast the game, but the English service of Radio Luxembourg covered it, and was followed by thousands of listeners in Sweden. See Ken Brooks, *Ingemar Johansson: Swedish heavyweight Boxing Champion* (Jefferson: McFarland & Company, 2016), p. 110, and Minutes of the "Comité de direction", 17th October 1959, RTL Group Archives, Luxembourg.

quoted by Maréchal.[24] This led the board of directors to consider, on at least two occasions, the possibility of extending their activities to the Scandinavian market, with a focus on Sweden. In 1959,[25] right after the match, and again in 1971.[26] As Radio Luxembourg already had listeners in the region, this would have integrated the margin as part of the main area covered. This implicated the possibility of evolution of this space, through the potential integration of a margin into the commercial interest, hinting at the flexibility of the entire ensemble.

One occasion during which Radio Luxembourg was involved in broadcasting outside of its usual markets was with Italy. Sources reveal that the Italian government and the R.A.I. (the *Radiotelevisione italiana*; the Italian public broadcaster) were negotiating with Radio Luxembourg over the possibility of buying airtime in the evenings to broadcast a programme in Italian. This programme was bought over Flemish airtime and the initial negotiation argued for a timeslot somewhere between half an hour to an hour and a half. [27] The final agreement, however, settled for a 15-minute programme at 7.30 pm, that was broadcast, daily, on medium wave, from 1st September 1972,[28] and it is not known when the collaboration ended. Interestingly, the programme was made by the R.A.I., and the Italian government paid Radio Luxembourg to broadcast it,[29] suggesting a collaboration including both private broadcaster and public institutions.

A third example shows interesting connections between Radio Luxembourg and migration movements to and from Spain. A short programme was planned to be broadcast on Radio Luxembourg from 1966, every

[24] Maréchal, *RTL. Histoire d'une radio populaire*, p. 183.

[25] Minutes of the "Comité de direction", 17th October 1959, RTL Group Archives, Luxembourg.

[26] Minutes of the "Comité de direction", 23rd April 1971, RTL Group Archives, Luxembourg.

[27] There is an inconsistency in the minutes that mention both ½ hours and 1,5 hours. Minutes of the "Comité de direction", 2nd July 1969, RTL Group Archives, Luxembourg.

[28] Minutes of the "Comité de direction", 26th September 1972, RTL Group Archives, Luxembourg.

[29] Minutes of the "Comité de direction", 23rd April 1971, RTL Group Archives, Luxembourg.

Sunday, at 8.30am, for Spanish workers in the Grand Duchy.[30] This programme was part of the shows for the audience in Luxembourg, and was made of local news of interest to foreign workers, completed by Spanish music. Part of the programme was made by using tapes sent from Spain. These tapes were only used for music, therefore indicating a probable fear of the programme being used for purposes other than entertainment and local news. Another connection between the station and Spain appeared in 1967, when staff members of the station went to host, on unnamed Spanish radio stations, programmes in French, targeting tourists spending their holidays there.[31] Little more than brief mentions could be found about both examples, nevertheless, they reveal an interesting side of the commercial station. Migration movements—a highly transnational phenomenon—had an impact on programmes by Radio Luxembourg, which appears to have modified them based on people crossing borders. The case of Spain developed above is particularly interesting, as it shows two highly different aspects of the station, depending on the group targeted by its programmes. On one hand, tourists spending their holidays in Spain were likely a loss for Radio Luxembourg, as they could not tune in to the station. This agreement with Spanish stations was probably a way to reduce the impact of summer holidays on listenership, and it echoes commercial concerns. On the other hand, the programme dedicated to Spanish workers in Luxembourg followed a different purpose. It was part of the *programmes luxembourgeois*, which were born of agreements with the government of the Grand Duchy, which wanted national programmes in exchange for letting the station broadcast from its territory. In this regard, the station operated what can be described as a public service mission, much like many public broadcasters in the neighbouring countries were doing. Moreover, offering programmes for a group of workers and migrants reveals an aspect of the station that makes it closer to community radio, suggesting its adaptability to diverse purposes.[32] In both cases, these examples infer a connection

[30] Minutes of the "Commission des programmes", 7th January 1966, RTL Group Archives, Luxembourg.

[31] Minutes of the "Comité de direction", 20th June 1967, RTL Group Archives, Luxembourg.

[32] It was not uncommon for radio stations in the 1960s to host programmes catering for the needs and interests of foreign workers. For the West German and British examples, see C. Hilgert, A. L. Just & G. Khamkar, "Airtime for Newcomers. Radio for Migrants

between the transnationality of broadcasting and the transnationality of migration.

The last example slightly differs from the other ones, as it concerns a programming possibility that never actually took place. There were indeed lengthy discussions in Radio Luxembourg about the possibility of broadcasting a programme in Greek, and these discussions and exchanges reveal a particularly interesting episode of the station's international history. In September 1967, a representative of the Greek Broadcasting Institute contacted Radio Luxembourg to begin negotiations regarding the broadcast of a daily half-hour long programme, in Greek, on medium wave, for Greek workers living in Germany.[33] The first letter included potential scripts, accompanied by translations in French. While the initial response by Radio Luxembourg was a positive one,[34] the Luxembourgish government, however, declined the offer. The refusal was due to Greek not being one of the languages spoken by the regular audience of Radio Luxembourg.[35] This echoes the case of Spanish workers, who were living in Luxembourg, and therefore could be targeted by such programmes for migrants. In both cases, the projects received a form of support from governments in their home countries. Interestingly, even though the project fell through, it seems that Radio Luxembourg later collaborated, nevertheless, with the Greek government to help them re-organise their audiovisual institutions.[36]

These examples showcase different scenarios of adaptability of the broadcasting space. In the case of the collaboration with the R.A.I., it is a broadcast to another country for the local audience, exchanging airtime dedicated to Flemish programmes in an interesting, short-lived, public–private partnership. The second case is fascinating as the station

in the United Kingdom and West Germany, 1960s–1980s", *Media History* (26:1, 2020), pp. 62–74.

[33] Correspondence from B. Aslanidis to Radio Luxembourg, 16th September 1967, RTL group Archives, Luxembourg.

[34] Correspondence from G. Graas to D. Chronopoulos, 26th September 1967, RTL Group Archives, Luxembourg.

[35] Correspondence from the *Commissaire du gouvernement* to the C.L.T., 7th October 1967, RTL Group Archives, Luxembourg.

[36] Minutes of the "Comité de direction", 1st March 1968, RTL Group Archives, Luxembourg.

Minutes of the "Assemblée générale annuelle", 21st May 1968, RTL Group Archives, Luxembourg.

catered to the needs of its audience, compensating the loss of listeners due to summer holidays, by simply having hosts follow them to Spain. The second and third examples display a different dynamic, as a foreign government asked Radio Luxembourg to cater for the needs of citizens abroad, raising the question of the mission of the station and its relationship with Luxembourg. The three cases display various facets of the complex framework in which Radio Luxembourg evolved, between monetary gain, mission, and oversight of the Grand Duchy. In relation to the transnational space, these examples point to the complexity of the nature of the actors involved (*i.e.* public, governmental, private and linked to migrant communities).

The analysis conducted in this chapter has focused on the construction of the commercial radio space through its edges. By looking at the core of radio production—the place where radio happened—as well as the more marginal areas covered by the transmitters, this chapter has shown the existence of cross-country flows and dynamics as a structural feature of the studied space. Moreover, the question of margins reinforces the understanding of the Iron Curtain as a permeable border, not only for politically oriented stations but also for commercial actors such as Radio Luxembourg. The point on short-lived attempts raises the existence of a European broadcasting space. And, as Suzanne Lommers showed, the existence of such a space is a phenomenon that can be traced back to the interwar period. Europe was then envisaged as 'a world of brotherhood between nation states',[37] a model supported by institutions such as the IBU. This model, however, was already challenged by stations like Radio Luxembourg, which had transnational ambitions. In the period studied here, commercial stations also developed a transnational space, one with peripheries and margins regularly challenged and questioned. This means that flexibility was also a key feature of the *dispositif* of commercial radio stations, particularly in regard to the broadcasting space.

References

Abdel-Massih, Milena and Anna Malewska-Szałygin. *Remembering Radio Luxembourg in the People's Republic of Poland*. Warsaw: Coma2, 2012.

[37] Suzanne Lommers, *Europe—On Air. Interwar Projects for Radio Broadcasting* (Amsterdam: Amsterdam University Press, 2012), p. 74.

Badenoch, Alexander, Andreas Fickers and Christian Henrich-Franke (eds.). *Airy Curtains in the European Ether: Broadcasting and the Cold War*. Baden-Baden: Nomos, 2013.

Beccarelli, Marine. "Micros de nuit: Histoire de la radio nocturne en France, 1945–2012". PhD thesis., Université Paris-1 Panthéon Sorbonne, 2016.

Berg, Katja. *Grenzenlose Unterhaltung. Radio Luxemburg in der Bundesrepublik, 1957–1980*. Göttingen: Wallstein, 2021.

Berg, Katja and Anna Jehle. "'Through the Air to Anywhere': Radio Luxembourg—A Transnational Broadcasting Station?" In *Transnational Broadcasting in Europe 1945–1990*, edited by Christoph Classen, 23–44. Frankfurt am Main: Peter Lang, 2016.

Brooks, Ken. *Ingemar Johansson: Swedish heavyweight Boxing Champion*. Jefferson: McFarland & Company, 2016.

Bull, Michael. "Sound, Proximity, and Distance in Western Experience". In *Hearing Cultures. Essays on Sound, Listening and Modernity*, edited by Veit Erlmann, 173–190. Oxford: Berg, 2004.

Erlmann, Veit (ed.). *Hearing Cultures. Essays on Sound, Listening and Modernity*. Oxford: Berg, 2004.

Fickers, Andreas and Pascal Griset. *Communicating Europe: Technologies, Information, Events*. Basingstoke: Palgrave Macmillan, 2019.

Fickers, Andreas. "Looking East-Watching West? On the Asymmetrical Interdependencies of Cold War European Communication Spaces". In *Television Beyond and Across the Iron Curtain*, edited by Kirsten Bönker, Julia Obertreis and Svenn Grampp, 2–24. Newcastle upon Tyne: Cambridge Scholars Publishing, 2016.

Fickers, Andreas. "The Radio Station Scale: A Materialized European Event". In *Materializing Europe: Transnational Infrastructures and the Project of Europe*, edited by Alexander Badenoch and Andreas Fickers, 252–254. Basingstoke: Palgrave Macmillan, 2014.

Fickers, Andreas. "Visibly Audible: The radio dial as mediating interface". In *The Oxford Handbook of Sound Studies*, edited by Trevor Pinch and Karin Bijsterveld, 411–439. Oxford: Oxford University Press, 2012.

Fickers, Andreas. "Die Anfänge des kommerziellen Rundfunks im Saarland". In *Medienlandschaft Saar von 1945 bis in die Gegenwart. Band 1: Medien zwischen Demokratisierung und Kontrolle (1945–1955)*, edited by Clemens Zimmermann, Rainer Hudemann and Michael Kuderna, 241–308. München: R. Oldenbourg, 2010.

Hilgert, Christoph, Alina L. Just & Gloria Khamkar. "Airtime for Newcomers. Radio for Migrants in the United Kingdom and West Germany, 1960s–1980s". *Media History* 26:1 (2020): 62–74.

Hilmes, Michele. *Radio Voices: American Broadcasting, 1922–1952*. Minneapolis: University of Minnesota, 1997.

Hochscherf, Tobias, Richard Legay and Hedwig Wagner. "Radio Beyond Boundaries: An Introduction", *Historical Journal of Film, Radio and Television* 39:3 (2019): 431–438.

Iriye, Akira and Jean-Yves Saunier (eds.). *Palgrave Dictionary of Transnational History*. Basingtoke: Palgrave Macmillan, 2009.

Jehle, Anna. *Welle der Konsumgesellschaft: Radio Luxembourg in Frankreich, 1945–1975*. Göttingen: Wallstein, 2018.

Le Breton, David. *Du Silence*. Paris: Éditions Métailié, 2015.

Lommers, Suzanne. *Europe—On Air. Interwar Projects for Radio Broadcasting*. Amsterdam: Amsterdam University Press, 2012.

Maréchal, Denis. *RTL. Histoire d'une radio populaire: De Radio Luxembourg à RTL.fr*. Paris: Nouveau Monde Éditions, 2010.

Mattelart, Tristan. *Le Cheval de Troie audiovisuel: Le Rideau de Fer à l'épreuve des Radios et Télévisions Transfrontières*. Grenoble: Presses Universitaires de Grenoble, 1995.

Méadel, Cécile. *Histoire de la radio des années trente: du sans-filiste à l'auditeur*. Paris: Anthropos, 1994.

Moores, Shaun. *Interpreting Audiences: The Ethnography of Media Consumption*. London: Sage, 1993.

Prince, Tony and Jan Sestak. *The Royal Ruler and the Railway DJ: Autobiographies of Tony Prince and Jan Sestak*. Maplewood: DMC Publishing, 2017.

Schülke, Sophie. "Endgültiger Sendeschluss in Berus". *Saarbrücker Zeitung*, January 17th, 2020.

Street, Sean. *Sound at the Edge of Perception: The Aural Minutiae of Sand and other Worldly Murmurings*. Basingtoke: Palgrave Macmillan, 2019.

Taylor, Timothy Dean, Mark Katz and Tony Grajeda (eds.). *Music, Sound, and Technology in America: A Documentary History of Early Phonograph, Cinema, and Radio*. Durham: Duke University Press, 2012.

CHAPTER 3

A Space Formed of Three Layers

Both Europe n°1 and Radio Luxembourg were conscious of their transnational reach, and the space studied in this chapter can be sliced into three layers, resulting from three different aspects of the broadcasters and how they considered it. These three layers—technical, commercial, and imagined—are not independent of one another, and there are various connections between them. For instance, how well a station was heard in a region had a direct impact on the potential size of the market in said region. Both the first and second layers had, in return, an impact on the third layer of the transnational space, the imagined one, as it was building on the achievements of the first two. It is through their influence on these three layers that commercial radio stations appropriated the transnational space in which they operated, as part of a radio *dispositif* deployed by commercial stations.

1 The Technical Layer

The first so-called layer studied in this chapter is a technical one; a layer focusing on radio waves and the stations' power to broadcast to a large transnational space, therefore enabling the two other layers studied in this chapter. In this layer, transmitter sites are at the centre of the space

R. Legay, *The Cultural Histories of Radio Luxembourg and Europe n°1*,
https://doi.org/10.1007/978-3-031-46250-4_3

studied as the very origin of radio waves.[1] The very nature of radio is further reflected in this layer as it focuses on the airwaves and the technological power to be heard properly.[2] In return, it is the signal emitted by the transmitters which binds the technical layer; where it ends and so does the first layer. Commercial radio stations were highly dependent on the spaces they could reach, and needed a good understanding of them, as well as those of their competitors'. The perspective of Europe n°1 is central in the following pages, due to the possibility by the author to consult maps and working documents made at the station's transmitter. These archives reflect the evolution of the transmitter and its power, the zones the airwaves could reach in various European countries, and the varying degrees of sound quality as the waves travelled. In this layer, the relationship between stations and space is strongly rooted in the transmitters, meaning that the layer is not a uniform one, as distance, direction, and physical obstacles have an impact on the quality of broadcasting.

The Importance of Maps. To appropriate this transnational space, commercial radio stations engaged in the creation of maps to visualise the reach of their transmitters. By doing so, they could obtain a visual representation and a better understanding of the spaces in which they could broadcast. This played a particularly interesting role in relation to technical evolutions and competition with other stations. Maps were powerful tools to appropriate this space, as they mix technical data, such as the power of the transmitter, with geographical information about relevant territories (*e.g.* their main cities). By extension, such maps are wonderful sources for historians. As mentioned, the focus in this section is only from the perspective of Europe n°1, due to the possibility of consulting maps made at its transmitter's site.

Measuring, understanding, and mapping Europe n°1's reach appears to have been in progress since the early days of the station, according to a series of maps, made for internal use, found in the station's archives.[3] The

[1] Therefore, the image carried by the term 'broadcasting' is particularly well illustrated in this layer. To broadcast originally meant "to sow by hand, in semi-circles, as wide as possible", and carries the idea of a central bucket, sending seeds to places where they are lacking. Jostein Gripsrud, *Understanding Media Culture* (London: Arnold, 2002), p. 262.

[2] What defines properly is obviously subject to discussion, and, unfortunately, the sources do not inform about what was meant by that at the time.

[3] Map of the 'Courbes de propagation', Europe n°1, June 1955, C^{2}DH archives, Felsberg transmitter site collection.

earliest source found among the maps, indeed, dates back to 1955 (see Fig. 1), and shows the radio waves of the transmitter and their potential to be heard throughout mainland France, and most of Western Europe. This map illustrates key elements that appear throughout the sources and have an impact on the technical layer. The direction of the Felsberg transmitter was roughly towards Bordeaux, in France, meaning that regions close to the transmitter but in an opposite direction were poorly reached by the station. This is the case, for instance, with Karlsruhe and Orléans. The signal was roughly the same around both cities, even though Karlsruhe is about 170 kilometres away from the transmitter, and Orléans, on the other hand, is 500 kilometres away. Therefore, there was some sort of geographical distortion in the technical layer that influenced distances, alongside the fact that the signal strength decreases the further away it is from the transmitter.

In a map from February 1974 (see Fig. 2), one can read the perception Europe n°1 had of its listening zones depending on its quality; more precisely, the station defined two zones of 'comfort'.[4] The first zone, deemed 'very comfortable', engulfed most of mainland France,[5] with the exception of Nice and its surroundings, on the Mediterranean coast by the Italian border, and of the Pyrenean region, in a stretch of land from the Atlantic coast to the Mediterranean one, and which includes Perpignan, although Pau and Tarbes belong in this first zone. This zone includes Luxembourg, most of Belgium (with the exception of the most Northern part), Maastricht and its surroundings, a part of East Sussex; more precisely, a stretch of land going roughly from Eastbourne to Hastings. The zone also includes most of the French-speaking part of Switzerland, Bern, and part of the Mittelland, as well as the most Western part of the Piedmont region in Italy. The map is far less precise in Germany, especially around its point of origin, due to the fact it was mostly targeting France. It appears, however, that the Saarland, most of the Rheinland-Pfalz, and the Western part of the Baden-Württemberg, including Freiburg, Baden-Baden, and Karlsruhe, were part of the zone for 'ideal' listening to Europe n°1. The second zone is for a 'comfortable' listening of the station, and includes the rest of metropolitan France

[4] Map of the 'Schéma des zones d'écoute', Europe n°1, 15th February 1974, C[2]DH archives, Felsberg transmitter site collection.

[5] Understood here as mainland France, therefore, not including Corsica and overseas territories.

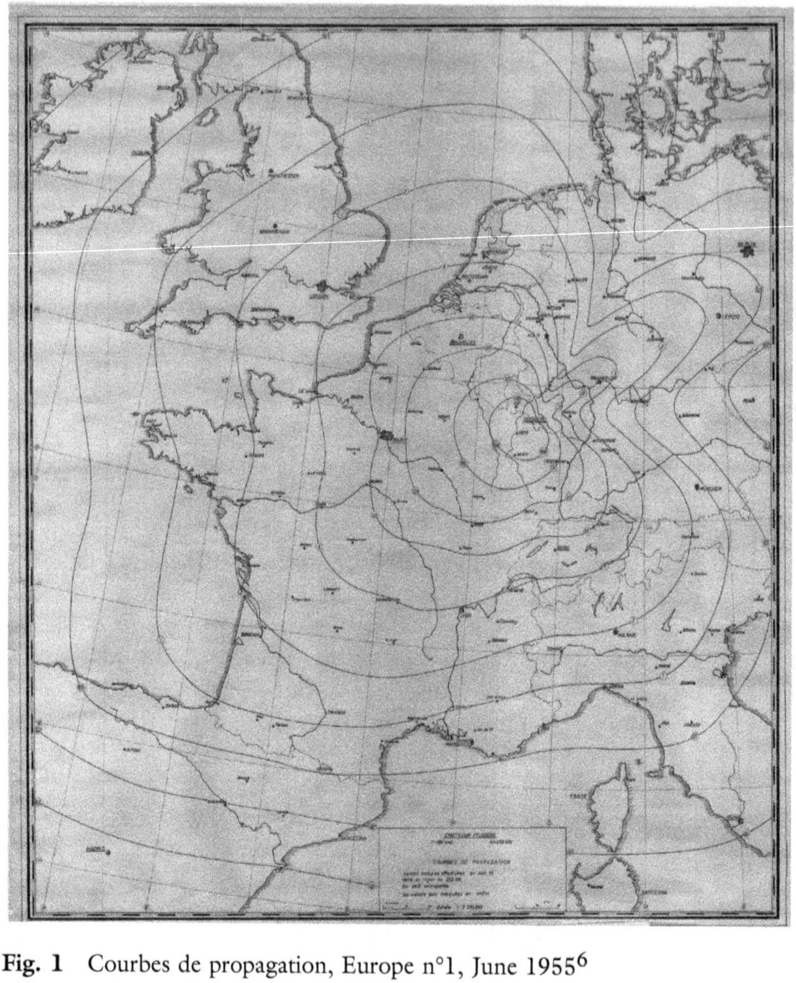

Fig. 1 Courbes de propagation, Europe n°1, June 1955[6]

(although not Menton and the Principality of Monaco), a stretch of land in the North-East of Spain, covering the region along the Pyrenees (part of the regions of Catalonia, Aragon, Navarre, and the Basque Country),

[6] Map of the 'Courbes de propagation', Europe n°1, June 1955, C²DH archives, Felsberg transmitter site collection.

the Northern end of Belgium and the Southern part of the Netherlands, roughly to Eindhoven, Rotterdam, and The Hague. This second zone reaches quite far North in Britain, including an area stretching from slightly south of Norwich to the Southern part of Wales, including cities such as Cambridge, Oxford, Bristol, Cardiff, and Swansea, and all others South of them, including London. It seems important to detail this space as it represents the height of Europe n°1's reach at the end of the period. Unfortunately, it is not possible to know what these two zones truly entailed, and on what the distinction was based, however, it points to a refinement of the understanding of the transnational space. Moving on from more 'raw' information, such as in the 1955 map, to the question of comfort in listening in the 1974 map infers that other factors were being taken into account, such as the listeners' perspective.

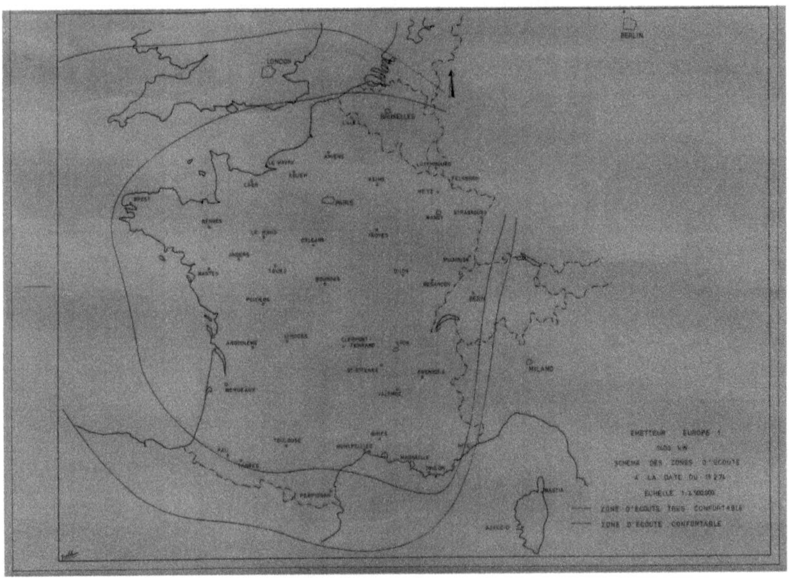

Fig. 2 Schéma des zones d'écoute, Europe n°1, February 1974[7]

[7] Map of the 'Schéma des zones d'écoute', Europe n°1, 15th February 1974, C²DH archives, Felsberg transmitter site collection.

A closer look at some of the maps which measured the *feldstärken* (field intensity) of Europe n°1's transmitter reveals the technical evolution of the broadcaster, from its early days throughout the Sixties. More precisely, these maps show the strength increase of the electric field, expressed by concentric waves, starting from the Feldsberg transmitter in Saarland, each measuring the strength in megavolt/metre (mV/m). Naturally, each wave strength decreases when compared to the one closer to the broadcasting site. An overview of this evolution in technical strength can be made by looking at maps made in four different years (1955, 1962, 1964 and 1971)[8] and four different cities in France (La Rochelle, Brest, Lille & Strasbourg). The first two cities are relatively far from the transmitter, on the West coast of France, while the other two are closer, respectively in Northern and Eastern France, therefore they give a reasonable overview of the French metropolitan territory. According to the information on the maps, the following table (see Fig. 3) reports that the measurement was made.

These numbers clearly indicate a strong technical evolution of Europe n°1, and how it reinforced its broadcasting power, particularly in its first

	1955	**1962**	**1964**	**1971**
La Rochelle	< 2 mV/m	6 mV/m	7 mV/m	7 mV/m
Brest	> 1 mV/m	3 mV/m	3,5 mV/m	3,5 mV/m
Lille	15 mV/m	17,5 mV/m	< 20 mV/m	< 20 mV/m
Strasbourg	35 mV/m	< 50 mV/m	75 mV/m	75 mV/m

Fig. 3 Strength of the Saarland transmitter in 4 French cities (in mV/m)

[8] Map of the 'Courbes de propagation', Europe n°1, June 1955, C²DH archives, Felsberg transmitter site collection.

Map of the 'Courbes de l'intensité de champ', Europe n°1, 25th November 1962, C²DH archives, Felsberg transmitter site collection.

Map of the 'Courbes de l'intensité de champ', Europe n°1, 1st April 1964, C²DH archives, Felsberg transmitter site collection.

Map of the 'Courbes de l'intensité de champ', Europe n°1, November 1971, C²DH archives, Felsberg transmitter site collection.

decade. This technical evolution of the station appears in the minutes of Radio Luxembourg's board of directors meetings too. They mentioned works undertaken in Saarland, while mentioning their own amelioration of their transmitters, referencing this competition and evolution in parallel to the commercial radio stations.[9]

The figures above also reveal important disparities between the four cities, depending mostly on the distance from the transmitter. Other factors, appearing clearly through a geographical lens and shown on the maps, also played a part in shaping these disparities. As the focus is on France, some mountain chains (the Jura, the Massif Central and the Alps) appear as disruptive elements for radio wave propagation, and particularly the Vosges, which interfered with broadcast to Alsace.[10] Another factor to take into account is the difference between night and day for broadcasting, something that was well known to Felsberg technicians, as revealed by the existence of maps measuring the signal's strength at night and during the day.[11] The orientation of the transmitter also influenced the areas reached by the stations. The comparison to be made here concerns Europe n°1 and the French service of Radio Luxembourg, as both transmitters were oriented 'roughly' in the same way: towards France. It is, however, crucial to insist on the adjective 'roughly', as demonstrated in one of the maps, they were not pointed exactly in the same direction, meaning, for instance, that RTL's broadcast to Belgium was better than its competitors, while it was inferior in the South-West of France. This difference echoes Radio Luxembourg's interest in broadcasting to Belgium, and also was the reason why the station tried to balance this inferiority in that region. One instance can be found in attempts to reach out to Radio Andorre and to use their transmitter as

[9] More precisely, the board of directors discusses on regular occasions the amount of budget allocated to the repairs and amelioration of their transmitters and antennas.

[10] Sources found in RTL Group Archives show a couple of simplified maps comparing the orientation of the transmitters of Europe n°1 and of RTL. Combined to the presence of mountain chains in France (mostly the Massif Central, the Alps, the Jura and the Vosges), these maps attest of a difficult presence of both stations in Southern France, due to the combination of these two factors. See "Principaux résultats de la 4ème vague de l'enquête C.E.S.P. 1971", 1971, RTL Group Archives, Luxembourg.

[11] Map of the 'Störabstand Linien (Tages und Nacht)', Europe n°1, 8th October 1959, C²DH archives, Felsberg transmitter site collection.

a relay for Radio Luxembourg's programmes.[12] Due to the location of the station in the Pyrenees, it would have been the perfect way to broadcast to the area otherwise neglected by the orientation of the transmitter. According to sources, however, this partnership never happened as Radio Andorre was simply not interested.[13]

La guerre des ondes. Sources reveal that both stations were keeping an eye on each other, comparing their reach and strengths in France through maps.[14] This hints at a wider *guerre des ondes*—war of airwaves—which once again placed technical characteristics as a central factor of the stations' competition. These maps are therefore a visual and geographical representation of this relationship. Other stations were monitored according to these technical concerns, such as Radio Monte Carlo,[15] however, Radio Luxembourg and Europe n°1 were the main opponents.

It is tempting, therefore, to see these maps as tools for a commercial war in this regard. The image might appear slightly extreme, but it would be in keeping with the primary purpose of Geography—war—according to Yves Lacoste. In his controversial book, first published in 1976, the French geographer developed a vision of the discipline, understood as the 'methodological description of spaces, from so-called "physical" aspects to economic, social, demographic, and political characteristics', as to be used, primarily, to wage war. This fact, described as something easily forgotten and masked by contemporary images of the discipline[16] is nevertheless portrayed as essential. Beyond the provocative title, Yves Lacoste makes numerous valid points, particularly regarding maps. They are presented as geographical representations *par excellence* and as tools of

[12] Hinting at another example of the station trying to change and improve its broadcasting space.

[13] Minutes of the "Comité de direction", 9th January 1965, RTL Group Archives, Luxembourg.

[14] Map of the 'Feldstärken von Radio Luxembourg' (N-E of France), Europe n°1, December 1962, C^{2}DH archives, Felsberg transmitter site collection. Map of the 'Courbes des intensités de champs diurnes' (Europe 1 & RTL), Europe n°1, 26th January 1976, C^{2}DH archives, Felsberg transmitter site collection. Map of the 'Courbes des intensités de champs diurnes' (RTL), Europe n°1, 26th January 1976, C^{2}DH archives, Felsberg transmitter site collection.

[15] Map of the 'Feldstärken (Tag) von RMC', Europe n°1, 25th March 1976, C^{2}DH archives, Felsberg transmitter site collection.

[16] Referring to France, he explains geography is often portrayed as useless and boring, due to the way it is taught in the country.

dominance and strategy, over a space and the people living in it.[17] In the case of the *dispositif* of commercial radio stations, these elements appear highly relevant. Physical geography, especially mountains, is essential as it impacts the propagation of radio waves, while human geography informs about clusters of potential listeners, and political considerations are key to the nature of the *radios périphériques*.[18] Needless to say, the 'war' between Radio Luxembourg and Europe n°1 has always been a metaphorical one and that is how the term is understood here. Both stations engaged in this competition, and historical maps found in the archives of RTL Group show that they also used maps as a tool to compare the two commercial stations' presence in France.[19] Not only are all these aspects of geography present on the maps, so too are the competitors, whose influence was also studied with care, influencing the importance of strategic decisions. As Lacoste points out, geography and, by extension, maps are not as 'harmless' as one could think, but integral to understanding the crucial impact of the technical layer; in particular, regarding the station's relationship with one another.

The *guerre des ondes* in French, a term commonly used to describe conflicts between broadcasters, often refers to political and propaganda questions. Alongside the notorious Voice of America and Radio Free Europe, one could add the BBC German East Zone Programme, broadcasting to the DDR, which faced opposition from the Berliner Rundfunk and Moscow Radio.[20] It is important to make this distinction with commercial broadcasters, however, it is still relevant to use the term war.

Mapping the Margins. Some of the maps made at Europe n°1's transmitter site offer insights into the station's attempt to map the edges of its

[17] Yves Lacoste, *La Géographie ça sert, d'abord, à faire la guerre*, 2nd ed. (Paris: La Découverte, 1985), p. 11.

[18] As previously mentioned on a few occasions, commercial radio stations were founded in Luxembourg and in the Saarland partly to avoid State monopoly on broadcasting in neighbouring countries.

[19] The maps, however, are not of the same quality. The information displayed on the maps (strength of the signal, quality of reception, etc.) were added by hand, and in basic figures, probably because the maps were supposed to be simple visual aids, rather the actual original documents. See "Principaux résultats de la 4ème vague de l'enquête C.E.S.P. 1971", 1971, RTL Group Archives, Luxembourg.

[20] Emily Oliver, "A Voice for East Germany: Developing the BBC German Service's East Zone Programme", *Historical Journal of Film, Radio & Television* (39–3, 2019), pp. 568–583.

transnational broadcasting space. This attempt can be seen in maps which offer information about other stations in those territories, as well as maps measuring Europe n°1's reach outside of the French market. A series of three maps from 1978, for instance, shows how the station could reach the German Democratic Republic,[21] and they reveal that Europe n°1 was aware of its broadcasting potential across the Iron Curtain. Two maps show the station also had interests in stations from Yugoslavia broadcasting to Italy,[22] pointing at an eagerness to understand more than the station's own reach, by integrating foreign actors.

Although it does not bear any date, one map shows Europe n°1's interest in its reach to the United Kingdom and to the Republic of Ireland,[23] likely around the same period as the other maps. The strength of the signal was not particularly strong (roughly 5 m/Vm around Portsmouth and London for instance, and it gradually decreased North of them, and close to zero in Scotland). Contrary to Radio Luxembourg, Europe n°1 was not broadcasting in English, therefore the British Isles did not offer a commercial interest, but this map still reveals two elements. First, Europe n°1 could, theoretically, be heard in the United Kingdom and Ireland. Second, this map indicates the station wanted to know its technical presence in this region, in order to obtain a global understanding of its European reach. Even though Belgium was not as prominent to Europe n°1 as it was for Radio Luxembourg, the Saarland-based station still measured its transmitters' strength to the country.[24] Interestingly, the map did not differentiate the various regions

[21] Maps of the 'Propagation vers la RDA', Europe n°1, 24th October 1978, Felsberg transmitter site.

[22] Map of the 'Feldstärkelinien in Italien', Europe n°1, date unknown, Felsberg transmitter site. Map of the 'Courbes de l'intensité de champ' (Italy), Europe n°1, date unknown, Felsberg transmitter site. Unfortunately, the maps are not dated. However, it appears safe to assume they are from the late 1960s to the 1970s, due to the dates on other sources sharing similar characteristics, such as the type of paper, its quality, and the format of the map. Among all the maps found in the transmitter site, none was dated from after 1978.

[23] Map of the 'Feldstärken' (United Kingdom & Ireland), Europe n°1, date unknown, Felsberg transmitter site.

[24] Map of the 'Intensité de champ théorique' (Belgium), Europe n°1, 30th September 1963, Felsberg transmitter site.

of Belgium, and there was no focus on Wallonia,[25] but rather on Belgium as a whole. This infers an interest strongly focusing on technical aspects and on the potential reach of radio waves.

While Europe n°1's transmitter was not oriented to broadcast towards the Bundesrepublik, its waves could still technically be picked up by receivers in Germany. This orientation is key here. Even though the transmitter was located in the Saarland, it was oriented towards Bordeaux, meaning that the reach was rather low in Germany, despite the geographical proximity. The reach of Europe n°1 in West Germany was understood in two ways. First, a map focused on the German Democratic Republic, and showed the reach of the station in a roughly semicircular space, going up to a few cities, from Aachen to Köln, Frankfurt, Tübingen and Freiburg.[26] Second, a map focused specifically on the Saarland, offering more precise measurements of the station's reach in the small German Lander.[27] Though Germany as a whole was not a key market for Europe n°1, it was different in the Saarland, with which the station had many contacts, for historical and geographical reasons.

These examples highlight the efforts deployed by Europe n°1 to measure, map, and, therefore, appropriate the margins of its transnational space. These margins were part of the technical layer, however, they were not necessarily included in the commercial one, implicating a crucial difference between the two.

The technical layer of the transnational space in which commercial radio stations operate offers an understanding of Europe n°1 and Radio Luxembourg as broadcasters. Through this lens, their transnational nature appears strongly as national borders are overlooked to focus on information such as data and geographical obstacles. In this layer, technology is crucial, as it directly influences transmitters and their reach over time, revealing an evolution throughout the period. The question of margins is also present, however, they appear less fragmented as they did in the previous section, as, through this technical lens, they are always present, even if at the very end of the signal's reach, the ultimate boundary of

[25] If Europe n°1 was specifically broadcasting to Belgium, French-speaking listeners in Wallonia could have represented a potential market.

[26] Map of the 'Zone de service 1403 kHz' (West of BDR), Europe n°1, date unknown, Felsberg transmitter site.

[27] Map of the 'Kurven gleicher Feldstärken' (Saarland), Europe n°1, date unknown, Felsberg transmitter site.

the technical layer. This brings out another key characteristic of this layer, which is the fact that it is unified, although not uniform. Unified in the sense that the broadcasting space is a whole, from the point of origin to the end. Not uniform in the sense that the strength of the signal—and, by extension, the comfort of listening—is directly impacted by the natural geography, the direction of the transmitter and, therefore, distorting distances and representations of the space. This also influences the second layer, where the signal strength and the comfort of listening are crucial for the establishment of listenership. Moreover, through the technical layer, maps become essential tools, both then and now, and are particularly highlighting the broadcasting competition, the *guerre des ondes*, which occurred between the various radio stations of the period. This competition was particularly striking in the commercial layer, although it still found roots in the technical one. Furthermore, it could be argued that the appropriation of the transnational space went through an understanding of the European broadcasting space in a large sense, and one not solely centred on the station and its main markets.

2 The Commercial Layer

The second layer studied in this chapter focuses on the transnational space as a market, a space of both actual and potential consumption of the stations' programmes. The nature of commercial stations makes this layer a vital one as their audiences[28] are at the heart of their revenue streams. Therefore, the *radios périphériques* appear to have put a great deal of effort into the understanding and expansion of their listenerships, seen as a market to conquer. In this section, the previous technical layer still plays a role but is further deepened by socio-economical aspects that are often discussed by the directions of the stations and are at the heart of audience surveys. This market layer is the most competitive one, in which the different actors are attempting to expand their reach and presence. Languages, cultures, and political context represent elements that bind this layer, and so does the signal, a central feature of the previous one. It

[28] Therefore, the audience is conceived as a market, one of the three definitions suggested by Crisell, and this section uses a similar conception. In the vision of the audience as a market, they become actual or potential consumers of the stations' programmes. See Andrew Crisell, *Understanding Radio*, 2nd ed. (London: Routledge, 1994), p. 203.

is also a layer that is of specific importance to commercial broadcasters, separating them from many other radio stations of the period.

The Surveys. As mentioned previously, commercial radio stations kept a close eye on the evolution of audience shares and listenership, especially in France, as the country represented their largest market. To do so, they relied on surveys conducted by external groups which shared their findings with the multiple actors of the commercial broadcasting industry; radio stations of course, as well as sponsors and advertising agencies. The point of these surveys was to have a picture of radio consumption at the time, and one that was as clear as possible. However, the ways to do so were not always the same, and, therefore, from one year to another, from one institution behind the survey to another, elements taken into account varied. Despite these changes, some elements appear to have remained consistent, and they constitute a perfect entry point into these sources. These surveys also grew in scope after the first broadcast of Europe n°1 in 1955, and then again throughout the 1960s to become more precise and detailed.[29]

First, the radio stations concerned by the surveys always included Europe n°1, Radio Luxembourg—RTL from 1966 onwards—and the public broadcaster France Inter, as these three were the most listened-to stations in the Long Sixties in France. Sometimes, other stations, such as Radio Monte Carlo, would also be part of the inquiry, although this was rare.[30] In the following analysis, listenership of stations other than the three mentioned above was significantly lower, they are therefore considered negligible. Secondly, the surveys consulted always include data based on two types of audience measurement. One measurement is the share of listeners, expressed in percentage. It is a twofold factor. On one hand, it is the percentage of people, among those consulted, who listened to the radio, at any given moment of the previous day. On the other hand, it is the percentage among these listeners who listened to one of three main radio stations in France, at any given moment of the previous day. The other measurement is the Average Quarter-Hour, usually known as AQH, which measures the average number of broadcast radio listeners

[29] Anna Jehle, *Welle der Konsumgesellschaft: Radio Luxembourg in Frankreich 1045–1975* (Göttingen: Wallstein, 2017), p. 354.

[30] Information regarding audience shares for this commercial station is present in Auditoire global des stations de radio, Rapport de J. Daron, November 1965, Archives RTL Group, Luxembourg.

of a typical quarter-hour, expressed in percentage. It also included an average of quarter-hours for typical weekdays (Monday to Friday, 6am to 10.30pm). Some of the surveys consulted also included more details that could refer to listeners (*e.g.* gender, profession, and age), and some had geographical information: in which region listeners lived, and if they were from an urban or rural background.

These measurements were analysed together to rank the stations, based on their shares of listeners, and to understand how many people were listening to each station, for how long, and when the peak times were. In other words, it allowed a clearer understanding of the French broadcast market, by breaking down listenership hour by hour, and by dividing listeners into social groups. Therefore, surveys were used as a negotiation basis between the various commercial actors to set the prices of airtime for advertisement. Considering the importance of this stream of revenue for the *radios périphériques*, it is easy to understand how crucial the results were.[31] The following pages focus on these surveys, through the archives of RTL Group. They contain relevant sources, ranging from documents sent to the station by survey institutions, to internal reports made about these surveys, and minutes of meetings of the board of directors, during which surveys and reports were discussed.

These sources are useful in two ways, as two readings can be made of them. First, the surveys were supported by multiple institutions; sponsors and broadcasters, including Radio Luxembourg and Europe n°1. Due to these multiple supports, the surveys seem to have been accepted as the most objective offer by the various actors of the commercial broadcasting industry and therefore, are particularly useful for historical analysis. Surveys are, by nature, not an exact representation of the reality, of the radio market in this case. This representation was, however, the one used by the stations, and was a basis for many a decision, meaning it is of particular relevance to this work. Second, the sources available often mix original data from the surveys with reports, analysis, and comments from the commercial station, offering insights into Luxembourg's perception,

[31] It is also mentioned that advertisers are accustomed to changes in audience shares. The repercussions of a drop in listenership did not necessarily happen as soon as the surveys came out. One source goes even as far as talking about the "trust" of advertisers into Radio Luxembourg's capacity to adapt after bad numbers (see Minutes of the "Comité de direction", 16th February 1971, RTL Group Archives, Luxembourg), which relativises the impact of such surveys, that are to be analysed over a long period.

reflection, and strategy, as well as its understanding of its place within the French market and towards other stations.

There is unfortunately little access to audience measurement surveys for the entire period studied in this book, only for the years 1964 to 1971, with a gap in 1969. Moreover, the available survey for 1965 does not include the AQH. As mentioned previously, the content of the surveys evolves throughout the period, especially because they are not conducted by the same groups. In 1968, the C.E.S.P. (*Centre d'Etude des Supports de Publicité*) took over the surveys previously conducted by Ifop-Etmar,[32] and made them more precise. From 1970 onwards, the surveys were made in four phases per year,[33] and, according to Jacques Durand, were based on samples of 12,000 to 14,000 people.[34] In the 1965 survey, the document states that 2,211 people took part,[35] while 8,000 people were interviewed in 1966 and 5,000 in 1967,[36] which indicates an increasing accuracy of the results from 1966 onwards, and even more after 1968.

For more readability, the key figures extracted from the surveys and annex documents have been turned into a table (see Fig. 4) and a graph (see Fig. 5) available below. It is crucial to mention that the AQH from France Inter in 1971 has been corrected[37] from the percentage written in the original document.[38] As mentioned, there is no information available for 1969 and only partial information for 1965. Results for this year also appear slightly inconsistent with other years, therefore, it might be safer not to pay them too much attention. Based on these documents, it appears that readers of these surveys were aware of the slight increase

[32] Jacques Durand, "Les audiences de la radio", *Sociologie de la communication* (1:1, 1997), pp. 915–921.

[33] Minutes of the "Comité de direction", 2nd July 1970, RTL Group Archives, Luxembourg.

[34] Durand, "Les audiences de la radio", pp. 915–921.

[35] Auditoire global des stations de radio, Rapport de J. Daron, November 1965, Archives RTL Group, Luxembourg.

[36] Report on "écoute ondes longues", minutes of the "Comité de direction", 21st July 1967, RTL Group Archives, Luxembourg.

[37] The document mentioned a percentage of 3,15 as an average of the percentages obtained during the four waves of the year 1971 for France-Inter, which is simply incorrect. The correct percentage–3,40–was therefore written instead in the chart. After verification, this is the only instance of what appears to be a genuine mathematical error.

[38] "Principaux résultats de la 4ème vague de l'enquête C.E.S.P. 1971", 1971, RTL Group Archives, Luxembourg.

in radio listening overall between 1964 and 1971, with roughly 60% of the French population listening to radio on a daily basis. This increase was described, in 1970, as the biggest one of the last ten years,[39] implicating the 1960s as a decade of development for radio, and not regression, despite the popularisation of television.[40] Moreover, according to these surveys, Radio Luxembourg/RTL's audience share suffered from a sharp decrease at the beginning of the period, followed by more stability from 1966 onwards, at a share of roughly 20% of listeners. Europe n°1's share appears to have experienced a similar evolution to the other commercial radio, while always staying slightly ahead from 1966 onwards. Interestingly, this station—also the youngest of the three—started from the lowest share in 1964, before steadying a few per cent above Luxembourg. Overall, France Inter appears to have experienced a significant increase throughout the period, and an obvious lead from 1966 onwards, with a peak in 1968.

The year 1966 seems to be a particularly important one, as the three stations appear to have extremely close shares of listeners—something that is especially striking when looking at Fig. 5. It is also the year when some of the tendencies in radio listening in France seem to have stopped: Europe n°1's increase[41] stabilised, while Radio Luxembourg's fall stopped. This fact was clear in the mind of the latter, which wrote a lengthy report from the summer of 1967,[42] in which it is stated that the station was affected by a 2% loss of the share overall between 1966 and 1967, while the decrease was 28% between the numbers for 1964 and 1966. Similar figures regarding the AQH are stated, pointing at what could be called a 'stop of the haemorrhage' that affected Radio Luxembourg until 1966, the date at which it was renamed RTL and two key figures, Jean Prouvost and Jean Farran, took over the reins of the

[39] Minutes of the "Comité de direction", 2nd July 1970, RTL Group Archives, Luxembourg.

[40] In terms of production in France, for instance, the numbers for transistors are far higher than for television sets. See Elvina Fesneau, "Eléments pour une histoire du public des postes à transistors", *Le Temps des Médias* (2:3, 2004), pp. 118–125.

[41] If the results for 1965 are discarded.

[42] Report on "écoute ondes longues", minutes of the "Comité de direction", 21st July 1967, RTL Group Archives, Luxembourg.

	Overall		Radio Luxembourg / RTL		Europe n°1		France Inter	
	Annual share	AQH	Annual share	AQH	Annual share	AQH	Annual share	AQH
1964	58,40%		30,60%	4,50%	20,20%	2,71%	22,60%	2,13%
1965			26,50%		34,00%		32%	
1966	58,20%	11,50%	22,20%	3,40%	24,50%	3,90%	25%	3,56%
1967	62,20%	10,79%	21,70%	2,90%	24,10%	3,60%	29%	3,60%
1968	60,60%	11,03%	19,60%	2,97%	21,40%	2,97%	32,90%	4,43%
1969								
1970	61,60%	10,40%	20,80%	3,10%	22,70%	3,20%	27,90%	3,30%
1971	61,70%	10,60%	19,95%	2,85%	23,10%	2,85%	29,60%	3,40%

Fig. 4 Listening shares and AQH for the three main radio stations in France, 1964–1971

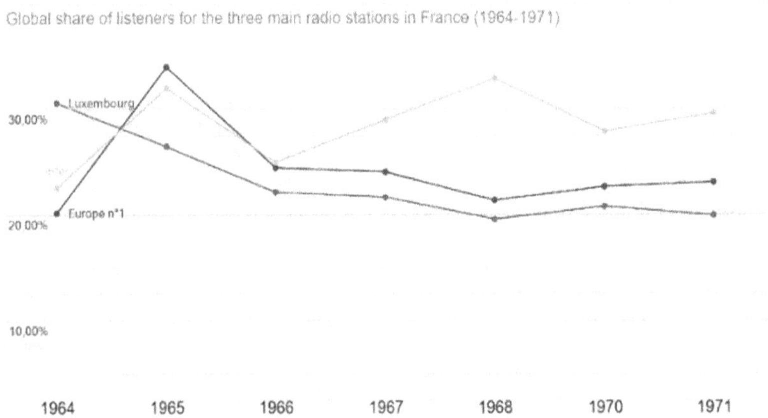

Fig. 5 Global share of listeners for the three main radio stations in France, 1964–1971

station.[43] Based on a 1967 survey and the previous ones—from 1966 and 1964—the newly appointed Jean Farran reckoned, in a 1967 meeting, that the fall in listening was contained, and, based on this, established further rules for the future programmes.[44] It is worth reflecting upon this moment of 1966 to show the importance of such surveys for a station like Radio Luxembourg/RTL, as they had a direct impact on the programmes broadcast, and, therefore, on the very sound of the station.

Who Were the Listeners? As mentioned previously, the surveys sometimes included a series of socio-geographical data regarding listeners. This information did not appear to play a function as prominent as the AQH and audience shares on the negotiation table. Radio Luxembourg appears, however, to have watched this information with care, as it helped the station understand the demographic dimension of its listenership. Not only how many people were listening to Radio Luxembourg/RTL mattered, but also, who was listening. The station and advertisers cared a great deal to know what kind of listeners were tuning it at which time. More precisely, the station appears to have been trying to win over specific types of listeners, those who were either young or had a high social status.[45] According to meetings of the board of directors in 1971, the station was pleased to have younger listeners and more listeners working in higher-up positions[46] than in the previous years,[47] however, they were

[43] More precisely, Farran took over the station, while Prouvost became the head of the CLT, the company behind Radio Luxembourg. Denis Maréchal, *RTL. Histoire d'une radio populaire* (Paris: Nouveau Monde, 2010), pp. 306–311.

[44] Minutes of the "Comité de direction", 21st July 1967, RTL Group Archives, Luxembourg.

[45] More precisely, and as Jehle showed, Radio Luxembourg/RTL—which identified itself as a family station in the 1950s and early 1960s—mostly considered its audience as a whole group, therefore attempted to focus on offering a "programme for all". Female listeners (or, in the station's perspective: housewives), however, were already targeted by more specific programmes, usually in the afternoon. The arrival of Ménie Grégoire and her team changed this perspective by offering a more complex and detailed vision of female listeners. See Jehle, *Welle der Konsumgesellschaft*, p. 248.

[46] In the original source, it is stated that there is an increase of listeners who are *cadres*, a French term to refer to people working as managers and executives. In other words, people with white-collar jobs and middle to upper-middle class lifestyle and income.

[47] Minutes of the "Comité de direction", 7th September 1971, RTL Group Archives, Luxembourg.

still from a lower social background than Europe n°1.[48] It can be argued that this quest for wealthier listeners had a direct link with advertisers, who would be looking for this kind of more lucrative market.

Questions regarding the status of listeners were directly linked with another key factor for Radio Luxembourg: its image. The importance of an international image is developed later in this book, and the focus here is on its more social aspects. Throughout the decade studied, Radio Luxembourg/RTL seems to have worked hard to achieve a change in the way it was perceived. This meant moving from a 'familial, serious, old-school and chatty'[49] station to a 'more modern, younger and more prestigious'[50] one, which was—according to the sources found at RTL Group—the prerogative of Europe n°1 until then. Two terms used appear particularly insightful: chattiness and prestige. The former refers to the amount of speech, which is seen in this context as an old-fashioned way of doing radio, while the latter is likely due to the wider radio landscape, in which commercial stations do not have the 'respectability' of public broadcasters, something which does not solely apply to France, but to other European countries. How the station managed to change its image is a complex set of dynamics, hinting at the mobilisation of the entire *dispositif* of the stations. It involved changes in programmes, sound and content of the station, and the effective promotion of these changes, especially through the press, with magazines and regional newspapers used to advertise news concerning the station.[51] In regards to changes, the year 1966 was, as mentioned, one of the most important moments.

The geographical location of listeners was also a concern, to a lesser extent, of Radio Luxembourg/RTL. It gave information regarding their social status and an urban population seems to have been the central focus of the station. The geographical location also helped shape the understanding of the so-called technical layer described previously. Identifying regions with more, or fewer, listeners also gave inputs for collaboration

[48] Minutes of the "Comité de direction", 3rd December 1971, RTL Group Archives, Luxembourg.

[49] Minutes of the "Comité de direction", 27th June 1966, RTL Group Archives, Luxembourg.

[50] Minutes of the "Comité de direction", 15th September 1970, RTL Group Archives, Luxembourg.

[51] Minutes of the "Comité de direction", 23rd April 1971, RTL Group Archives, Luxembourg.

with regional press, to promote the station in general or to push forward specific programmes for instance, depending on the expected outcome.

3 The Imagined Layer

The third layer of transnationalism concerns the self-image put out by the stations. They both used the fact that they were transnational actors to portray themselves as international broadcasters and institutions, building on the first two layers to shape their image. While the flows operating with the commercial stations were indeed transnational, they were translated, when disseminated, into different forms. Among the images spread by the stations, the idea that they were international regularly came back. Moreover, this internationalism can be refined, at times, as European and as global. This layer can be seen as more elusive, as it is, foremost, a representative one. It is, nevertheless, a particularly interesting one, fully integrated in the *dispositif* of commercial radio stations.

Portraying Their Audience as International. One-way commercial radio stations portrayed themselves as international was through their audience. Listeners and readers, in the case of magazines, were, at times, portrayed in a way that would create an international image. By process of metonymy, this would reflect on the stations, resulting in their development of a more international image. To understand how this portrayal occurred, this section focuses on examples of radio programmes which displayed this internationalism through their guests, as well as extracts from magazines, in which letters from readers contributed to the internationalism of the image.

One of the forms taken by the internationalisation of the audience by commercial radio stations can be found in the way they chose guests supposedly reflecting their listeners' opinions. In July 1963, Radio Luxembourg's journalist Jean Carlier[52] held a debate on the perception of Europe by young people.[53] The speaker presenting the news just before the debate introduced the segment by asking listeners about their opinion on the matter of Europe in the following way: 'you, French people, or

[52] Jean Carlier (1922–2011) was a French journalist, who worked for Radio Luxembourg, and RTL, for most of his career. He was the Director of Information of the station from 1967 to 1982.

[53] Journaux parlés de Radio Luxembourg et d'Europe n°1, 7th July 1963, Radio Luxembourg & Europe n°1, 01:02:50, Inathèque, Paris.

you, Germans, or Belgians, or Luxembourgish, or Europeans', creating a clear sense of an international audience.

The debate itself furthered this sense of internationalisation on air. Three French people, two men and a woman, were discussing, in French, with three Germans, two men and a woman as well. The first group were in studios in Paris, while the other three were in studios in Cologne, with Jean Carlier moderating the questions. After introducing themselves, the guests asked each other a series of questions regarding various political and memorial aspects. They discussed, among other things, the fear of Communism in Germany, the memory of war crimes in France by Nazis,[54] European integration and the possibility of the United States of Europe. Overall, the six young men and women appear to share similar pro-European perspectives and pacifist views.[55] Another similarity is the fact that the two women, who asked the first questions, were rapidly not heard during the rest of the debate.[56] Interestingly, Jean Carlier explains during the debate, which was edited before being broadcast, that he did not know what the questions would be, and often referred to the high level of expertise of the guests. They concluded their discussion by planning to meet in Cologne later that year, with the moderator explaining the station would help get them in touch.

This example reflects the various facets of the internationalisation of the audience by Radio Luxembourg. On one hand, listeners were directly addressed as French, German, etc. On the other hand, the guests were said to represent the diversity of the audience, and the content of the debate reflected this internationalisation. The broadcasting apparatus,

[54] As shown by Gerwarth and Malinowski, the violence generated by wars and conflicts, divisive events, plays a part in Europeanization, alongside more positive dynamics. While the young Franco-German guests were probably not born during World War II, they were likely more familiar with the Cold War and decolonisation conflicts, such as in Algeria. See R. Gerwarth & S. Malinowski, "Europeanization through Violence? War Experiences and the Making of Europe" in M. Conway & K. K. Patel (eds.), *Europeanization in the Twentieth Century: Historical Approaches* (Basingstoke: Palgrave Macmillan, 2010), pp. 189–209.

[55] One of the French men explains, for instance, that he came back from Algeria recently and had no intention of putting on a uniform again, which would be perceived, at the time, as a strong anti-militarist sentiment.

[56] Which echoes the long-lasting smaller presence of women's voices on radio. See Nozha Smati & Anne-Caroline Fiévet, "A la radio, la voix donne à écouter et à voir", 04/12/2017, INAglobal.fr. [http://www.inaglobal.fr/radio/article/la-radio-la-voix-donne-ecouter-et-voir-10010].

with studios in Paris and in Cologne, is also an example of transnational collaboration between France, Germany, and Luxembourg, from where the programme was broadcast. Moreover, the whole debate was about Franco-German relations and European integration and sentiment. The guests, identified as—and sound like—French and German, furthered this Europeanness, integrating this internationalisation in the audible fabric of the programme. As these guests were supposedly reflecting the audience, it is easy to assume that was how the station perceived and wanted to represent its listeners.[57]

Voices of Europeanisation. Among the images put out by the stations, the European one is likely the most carefully crafted. The choice of 'Europe n°1' as a name for a commercial radio station is a clear play 'with the European idea's symbolic capital and successfully promoted a vision of broadcasting as a transnational and cross-border medium',[58] therefore claiming this European calling and ambition. Radio Luxembourg was also, through its directors, carefully following the evolution of the European integration process. For instance, in 1960, they were pleased with the development of the Common Market, as they thought it would help the growth of the station.[59] They believed, it seems, that Radio Luxembourg, a 'European station'[60] was to benefit from the expansion of the European Common Market by extension.

The board of directors often discussed the Europeanness of the station. They insisted, for instance, that the French programme should reflect the 'international mission' of Radio Luxembourg, which was perceived to have a 'European calling'.[61] More precisely, this Europeanness was to

[57] From its first issue in 1962, the magazine *Salut les Copains*, published in collaboration with the eponymous radio show on Europe n°1, included pen pal sections for readers who wished to get in touch with each other. Here again, a sense of internationalism was developed.

[58] Andreas Fickers & Pascal Griset, *Communicating Europe. Technologies, Information, Events* (Basingtoke: Palgrave Macmillan, 2019), p. 224.

[59] Minutes of the "Comité de direction", 18th June 1960, RTL Group Archives, Luxembourg.

[60] Katja Berg & Anna Jehle, "'Through the Air to Anywhere': Radio Luxembourg—A Transnational Broadcasting Station?", in C. Classen (ed.), *Transnational broadcasting in Europe, 1945–1990* (Frankfurt am Main: PL Academic Research, 2016), pp. 23–44.

[61] Minutes of the "Comité de direction", 27th October 1962, RTL Group Archives, Luxembourg.

be expressed through news, and the editorial line was expected to follow this idea. The Head of Information for the French programme of Radio Luxembourg sent, in 1964, a letter to *all* journalists, reminding them of the European calling of the station.[62] This correspondence, which was also attached in an internal memorandum, is particularly insightful into the vision of the station and how it was to be transcribed in the programmes. The calling is perceived as a way to protect a form of objectivity against the various governments, especially the French one. The letter acknowledged a world in which a 'new and common humanity takes shape, despite races and borders', creating a perfect environment for radio to play a crucial role. The author continued by praising the power of radio and of Luxembourg, said to be at the 'crossroad of Europe' and an 'island of freedom'. These arguments were put forward to justify a particularly interesting injunction. All journalists, lucky enough to be picked by the station, were expected, despite their nationality, to renounce old loyalty and accept the fact that they belonged to the 'great homeland' of Europe.[63] It does not mean that journalists embraced, or even followed, the grandiloquent injunctions of this letter, however, it does reflect a particular vision the station had of itself. It also reflects a particular take on Europeanisation, one which goes hand-in-hand with European integration, the European Economic Community (EEC) and the European Coal and Steel Community (ECSC). Luxembourg played a key role in the ECSC as it was the home of its High Authority. Radio Luxembourg, based on the letter analysed above, portrayed itself as a haven for journalists who advocated an 'ideal or "dream" Europe', echoing research conducted on intellectual dissidents in Europeanisation.[64]

A regular concern of the board of directors was to carefully and diplomatically balance these different faces of Radio Luxembourg, depending on the various countries with which they had their activities. For instance, it was felt that the German programmes should not have a Frenchman

[62] Memorandum from Michel Moine to Jean Le Duc, 10 December 1964, RTL Group Archives, Luxembourg.

[63] Memorandum from Michel Moine to Jean Le Duc, 10 December 1964, RTL Group Archives, Luxembourg.

[64] J. Wardhaugh, R. Leiserowitz & C. Bailey, "Intellectual Dissidents and the Construction of European Spaces, 1918–1988" in M. Conway & K. K. Patel (eds.), *Europeanization in the Twentieth Century: Historical Approaches* (Basingtoke: Palgrave Macmillan, 2010), pp. 21–43.

at their head, in order not to antagonise potential German partners. In France, having a building in the heart of Paris to represent the station was deemed crucial, as it showed a diminution of the Luxembourgish identity there, by avoiding the mention of the CLT.[65] These observations show some of the challenges Radio Luxembourg faced to define and present itself as a European station.[66]

Grand Prix International. On one particular occasion, RTL invested in an important media event in order to promote its international and European image. *Grand Prix International*, an international musical competition, was organised in October 1969 for the first time. The countries involved were those covered by the station: France, Germany, Austria, Belgium, the Netherlands, Ireland, the United Kingdom, Switzerland (German and French-speaking regions) and, obviously, Luxembourg. Each country had a national competition to pick a song and artist,[67] who would then compete with other finalists for the prize. Listeners could make up their minds by listening to competing songs on the station's airwaves and cast their vote by sending postcards to the Villa Louvigny. The annual competition was only organised by Radio Luxembourg in 1969. Afterwards, the ZDF (*Zweites Deutsches Fernsehen*) took over on three more occasions. In 1973, the competition was abandoned.

Despite being rather short-lived, this international competition represents an interesting event for the study undertaken in this chapter. It was, first of all, a true transnational media event, relying on two types of cross-border flows: radio waves and postcards, with Luxembourg as an epicentre. Secondly, it showcased the implementation of Radio Luxembourg in many countries, building on its various audiences. Thirdly, the competition was envisaged as a way for the station to 'show to

[65] Minutes of the "Comité de direction", 26th September 1959, RTL Group Archives, Luxembourg.

[66] A concrete application of this European focus can be found in the trilingual broadcast of *International Night Programme*, simultaneously on short, medium and longwave, and with a presenter from three services (French, German and English) of Radio Luxembourg. The show was broadcast at night (midnight to 3am) for 3 years from 1961 onwards. Berg & Jehle, "Through the Air to Anywhere", in Classen, *Transnational broadcasting in Europe*, pp. 23–44.

[67] The artists had to be known in their country of origin, and the songs had to be original and written in one of the languages of their countries. Participants were also expected to send at least one other version of the songs in a different language, to be understood by many listeners.

the music world the leading position of Radio Luxembourg' and to 'prove the European influence of Radio Luxembourg to its listeners and advertisers'.[68] This shows how the competition built on the previous technical and commercial layers to project this transnational and European image, perceived as a sign of strength. Lastly, it could be argued that *Grand Prix International* was a media event, reinforcing its impact despite being short-lived. A media event is, for audiences, 'an invitation—even a command—to stop their daily routines and join in a holiday experience'[69] and can be defined further as 'certain situated, thickened, centring performances of mediated communication focused on a specific thematic core, cross different media products and reach a wide and diverse multiplicity of audiences and participants',[70] and this definition applies well to *Grand Prix International*. Likewise, Frank Bösch argued that the *Eurovision Song Contest* was a media event, and more precisely, a media-generated event.[71] Despite not being of the same importance as *Eurovision*,[72] *Grand Prix International* shared more than a few similarities with it, therefore supporting its classification as a media event.[73] In his definition of a media event, Bösch also insists on the importance of a

[68] Grand Prix International R.T.L., 22nd January 1969, RTL Group Archives, Luxembourg.

[69] D. Dayan & E. Katz, *Media Events: The Live Broadcasting of History* (Cambridge: Harvard University Press, 1992), p. 1.

[70] Nick Couldry and Andreas Hepp, "Introduction", in N. Couldry, A. Hepp & F. Krotz (eds.), *Media Events in a Global Age* (London: Routledge, 2010), pp. 1–20.

[71] Frank Bösch, "European Media Events", in *European History Online* (EGO), published by the Institute of European History (IEG), Mainz, 3rd December 2012. [http://ieg-ego.eu/en/threads/european-media/european-media-events/frank-boesch-european-media-events] (last consulted 23 January 2023). Moreover, European media events which have shaped a European Broadcasting space, have a long history, with roots up to the 1920s and 1930s, see A. Fickers & S. Lommers, "Eventing Europe: Broadcasting and the Mediated Performances of Europe", in A. Badenoch & A. Fickers (eds.), *Materializing Europe: Transnational infrastructures and the Project of Europe* (Basingstoke: Palgrave Macmillan, 2014), pp. 225–251.

[72] For instance, Christian Henrich-Franke coined *Eurovision Song Contest* as the "biggest transnational television event in Europe". See Christian Henrich-Franke, "Creating Transnationality through an International Organization? The European Broadcasting Union's (EBU) television programme activities", *Media History* (16:11, 2010), pp. 67–81.

[73] *Grand Prix International* could be seen in relation to *European Pop Jury*, another show featuring pop records which lasted two decades, which helps to map "the creation and fragmentation of technological and cultural spaces in Europe", see Alexander Badenoch, "'In what language do you like to sing best?' Placing popular music in broadcasting

physical presence, which coincides with *Grand Prix International* being held at Luxembourg's new City Theatre, where other broadcasters were invited to join to cover the event.[74]

It should be added that the stations played all competing songs—one by country—before the actual day of the competition, and the winning one was repeatedly played afterwards. This anchored the event for a longer period than just the day of the ceremony. *Grand Prix International* was furthermore present on the multiple programmes and languages of Radio Luxembourg, bringing together different spaces and audiences. All these elements therefore support the idea of this international musical competition playing a key role in weaving transnational and European images into the imagined layer of Radio Luxembourg. Part of the impact of this event also relied on the merging of the different spaces of Radio Luxembourg, which were usually divided by languages, except on this occasion.[75]

As seen previously, the appropriation of a transnational broadcasting space was a crucial stake for commercial radio stations, which relates to the 'Western project of the appropriation and control of space, place, and the other', described by Michael Bull.[76] Bull focuses on mobile sound technologies, such as portable radio and cassette players, nevertheless, his point is still relevant to describe the spatial appropriation process undertaken by the commercial broadcasters studied in this work. This process was analysed through three interconnected layers, representing three different lenses of analysis, and source material, and were bound by different elements. Nevertheless, as seen throughout this section, the

in post-war Europe", *European Review of History: Revue européenne d'histoire* (20:5, 2013), pp. 837–857.

[74] Grand Prix International R.T.L., 22nd January 1969, RTL Group Archives, Luxembourg.

[75] The event was also advertised in the pages of the *Fabulous* magazine, published weekly in collaboration with the English service of Radio Luxembourg, in the following way: "In fact it's going to be the festival of the year, folks (and FAb's Editor will be there, too, as a guest of 208)! The final fifteen productions will be played twice a day on all services of Radio Luxembourg for one week only, and a prize of a weekend for two in Luxembourg is being offered by 208 for the listener whose selection comes closest to that of the judges at the final." Fab 208 4th Oct 1969.

[76] Michael Bull, "Sound, Proximity, and Distance in Western Experience", in Veit Erlmann (ed.), *Hearing Cultures. Essays on Sound, Listening and Modernity* (Oxford: Berg, 2004), pp. 173–190.

layers were connected and built on one another. One key example can be found in a report from Radio Luxembourg and related to surveys. In there, a few pages meant for the board of directors present numbers regarding listenership and a series of quickly drawn maps of France,[77] illustrating key regions for the station and its competitors, in terms of signal and listenership. As such, these maps show the connections between the three layers, as they are based on the appropriation of a transnational broadcasting space through technical and commercial lenses, while maps are a representation of the space. Brought together, the three layers point to the existence of an ensemble of heterogeneous elements—as illustrated by the diversity of the layers—which are interconnected to form a *dispositif*. The appropriation of a transnational broadcasting space for commercial purposes echoes the 'dominant strategic function'.

References

Badenoch, Alexander. "'In what language do you like to sing best?' Placing popular music in broadcasting in post-war Europe". *European Review of History: Revue européenne d'histoire* 20:5 (2013): 837–857.

Berg, Katja and Anna Jehle. "'Through the Air to Anywhere': Radio Luxembourg—A Transnational Broadcasting Station?". In *Transnational Broadcasting in Europe 1945–1990*, edited by Christoph Classen, 23–44. Frankfurt am Main: Peter Lang, 2016.

Bösch, Frank. "European Media Events". *European History Online* (EGO). Institute of European History (IEG), Mainz. Online, 3rd December 2012. [http://ieg-ego.eu/en/threads/european-media/european-media-events/frank-boesch-european-media-events]

Bull, Michael. "Sound, Proximity, and Distance in Western Experience". In *Hearing Cultures. Essays on Sound, Listening and Modernity*, edited by Veit Erlmann, 173–190. Oxford: Berg, 2004.

Couldry, Nick, Andreas Hepp and Friedrich Krotz, eds. *Media Events in a Global Age*. London: Routledge, 2010.

Crisell, Andrew. *Understanding Radio*, 2nd ed. London: Routledge, 1994.

Dayan, Daniel and Elihu Katz. *Media Events: The Live Broadcasting of History*. Cambridge: Harvard University Press, 1992.

[77] The maps are clearly drawn by hand, and are quite simple, therefore serving as a visual support, rather than accurate presentation of data. See "Principaux résultats de la 4ème vague de l'enquête C.E.S.P. 1971", 1971, RTL Group Archives, Luxembourg.

Durand, Jacques. "Les audiences de la radio". *Sociologie de la communication* 1:1 (1997): 915–921.Gripsrud, Jostein. *Understanding Media Culture*. London: Arnold, 2002.

Fesneau, Elvina. "Eléments pour une histoire du public des postes à transistors". *Le Temps des Médias* 2:3 (2004): 118–125.

Fickers, Andreas and Pascal Griset. *Communicating Europe: Technologies, Information, Events*. Basingstoke: Palgrave Macmillan, 2019.

Fickers, Andreas and Suzanne Lommers. "Eventing Europe: Broadcasting and the Mediated Performances of Europe". In *Materializing Europe: Transnational infrastructures and the Project of Europe*, edited by Alexander Badenoch and Andreas Fickers, 225–251. Basingstoke: Palgrave Macmillan, 2014.

Gerwarth, Robert and Stephan Malinowski. "Europeanization through Violence? War Experiences and the Making of Europe". In *Europeanization in the Twentieth Century: Historical Approaches*, edited by Martin Conway and Kiran Klaus Patel, 189–209. Basingstoke: Palgrave Macmillan, 2010.

Henrich-Franke, Christian. "Creating Transnationality through an International Organization? The European Broadcasting Union's (EBU) television programme activities". *Media History* 16:11 (2010): 67–81.

Jehle, Anna. *Welle der Konsumgesellschaft: Radio Luxembourg in Frankreich, 1945–1975*. Göttingen: Wallstein Verlag, 2018.

Lacoste, Yves. *La Géographie ça sert, d'abord, à faire la guerre*, 2nd ed. Paris: La Découverte, 1985.

Maréchal, Denis. *RTL. Histoire d'une radio populaire: De Radio Luxembourg à RTL.fr*. Paris: Nouveau Monde Éditions, 2010.

Oliver, Emily. "A Voice for East Germany: Developing the BBC German Service's East Zone Programme". *Historical Journal of Film, Radio and Television* 39:3 (2019): 568–583.

Smati, Nozha and Anne-Caroline Fiévet, "A la radio, la voix donne à écouter et à voir", 04/12/2017, INAglobal.fr. [http://www.inaglobal.fr/radio/article/la-radio-la-voix-donne-ecouter-et-voir-10010]

Wardhaugh, Jessica, Ruth Leiserowitz and Christian Bailey. "Intellectual Dissidents and the Construction of European Spaces, 1918–1988". In *Europeanization in the Twentieth Century: Historical Approaches*, edited by Martin Conway and Kiran Klaus Patel, 21–43. Basingstoke: Palgrave Macmillan, 2010.

CHAPTER 4

Localism, Foreignness, Transnational Influences, and Competition

When researching commercial radio stations history, two aspects keep coming back: on one hand, they were putting out an international image, playing on a form of foreignness in their programmes and advertisement, while, on the other hand, they broadcasted a great number of shows highlighting local interests. These two aspects are therefore studied in this chapter, in order to both understand and reconcile them. Localism and globalism, in radio, have been studied and analysed by scholars.[1] The tensions between radio being global and local seem to be particularly strong from the late 1960s onwards, with the development of community radio, notably in the United Kingdom. In a similar manner to Guy Starkey, who wonders if radio was 'going global or turning parochial?',[2] one can bring this duality to a questioning of commercial radio stations. The stations studied in this book relied on an international image, as described in the previous chapter. Localism was, however, never absent from the programmes, as many shows offered a strong local anchor to their content. Localism is understood here as 'the specifically local

[1] See, for instance, "The Reflections on Tensions between Local and Global for the Radio Industry", in David Hendy (ed.), *Radio in the Global Age* (Cambridge: Polity Press, 2000), p. 60.

[2] Guy Starkey, *Local Radio, Going Global* (Basingstoke: Palgrave Macmillan, 2015), p. 163.

R. Legay, *The Cultural Histories of Radio Luxembourg and Europe n°1*,
https://doi.org/10.1007/978-3-031-46250-4_4

quality of some radio stations'.[3] This created a pendulum that constantly swung between these two elements of the stations, creating a particularly compelling and uncommon feature. The stations displayed, and played with, their international image, this idea of foreignness and, therefore, of liberty, freedom, and exoticism. They also cultivated their local quality, creating a fascinating duality of their identities. These two characteristics of localism and foreignness might appear contradictory at first glance; however, this chapter shows how commercial stations made it happen.

It is also crucial not to forget that commercial radio stations were part of a larger media landscape, in which various actors (governments, other stations, whether public, commercial, or offshore) all interacted with one another. This radio ecology[4] is a particularly rich and complex one, therefore, this chapter also sheds light on a few specific examples of transnational influences that occurred in the space in which commercial radio stations existed. These influences are dealt with in two phases. First, the foreign influences and connections, through three examples, and second, the complex relationship between Europe n°1 and Radio Luxembourg, made of competition, close surveillance, and occasional collaboration. This relationship forms a fundamental aspect of the transnational space studied in this chapter and the previous ones.

1 Foreignness, Liberty, Attractivity, and Localism on the Move

As mentioned, the image of foreignness developed by commercial radio stations is aligned with the idea of liberty. More precisely, a liberty that applied to news and editorial freedom, as well as entertainment programming and musical choices. These ideas were directly constructed in opposition to public broadcasters, and their—both actual and perceived—strict

[3] Hugh Chignell, *Key Concepts in Radio Studies* (London: Sage, 2009), p. 131.

[4] The idea of radio ecology, mentioned in Föllmer & Badenoch, is useful to portray the functioning of commercial radio stations, which actively engaged with various 'organisms' across countries. The metaphor is also helpful to develop the flows and evolutions that connect these actors together. Golo Föllmer & Alexander Badenoch (eds.), *Transnationalizing Radio Research: New Approaches to an Old Medium* (Bielefeld: Transcript, 2018), p. 12.

control over programmes and editorial lines, following a more 'paternalistic' mission.[5] The *British Broadcasting Corporation*, for instance, inherited a patriarchal and paternalist image, based on the Reithiasnism of its early days, with a strong high-culture stance. This image dates back to the pre-war period, but was still present in the Long Sixties, as notably shown by Martin Dibbs when studying an illustration for the Sunday programme of the *Radio Times*, an organ of the BBC, where members of a family are gathered together around the radio, which is exactly how listeners were portrayed before the Second World War.[6] The BBC actively worked to change this image in the 1960s, notably with the launch of BBC Radio One in 1967. Other changes in the following years went for 'arguably a more populist and digestible type of radio'.[7] Part of these changes were born of the realisation that the radio audience was increasingly moving towards television and Radio Luxembourg.[8] As a foreign broadcaster, the station had cultivated a different image.

Even though the question of news and editorial freedom plays a minor role in this book, it is a subject worth mentioning when looking at the stations' images. One striking example is France, where this was a particularly tense question in the Sixties. Throughout the decade, staff members of the public broadcasting agency showed defiance towards the censorship applied by the government. The *Ministre de l'Information* indeed had direct oversight over news content,[9] something long contested and which reached a breaking point in 1968. During the protests and riots

[5] Regarding the BBC paternalistic mission, see Su Holmes, *British Television and Film Culture in the 1950s* (Bristol: intellect Books, 2005), p. 114. In France, public service radio was strongly linked to General De Gaulle's image and relationship to radio, especially in war times. See Hélène Eck, *La Guerre des ondes. Histoire des radios de langue française pendant la Deuxième guerre mondiale* (Paris: Armand Colin, 1985), p. 134.

[6] Martin Dibbs, *Radio Fun and the BBC Variety Department, 1922–1967. Comedy and Popular Music on Air* (Basingstoke: Palgrave Macmillan, 2019), p. 229.

[7] Hugh Chignell, *Public Issue Radio: Talks, News and Current Affairs in the Twentieth Century* (Basingtoke: Palgrave Macmillan, 2011), p. 98.

[8] Dibbs, *Radio Fun and the BBC Variety Department*, p. 233.

[9] Alain Peyrefitte (1925–1999), who served as Minister for Information from 1962 to 1966, explained how he decided over the phone, everyday at 5 pm, the lines of the evening news on both television and radio. See Alain Peyrefitte, *Le Mal français* (Paris: Plon, 1976), p. 69.

of May 1968, a long-lasting strike of the ORTF[10] occurred, thus many listeners moved to both commercial stations to follow the news.[11] RTL and Europe n°1 did not suffer from this censorship,[12] therefore reinforcing the perceived links between foreignness and liberty. Based on both examples developed above, one can see that this image of perceived freedom is indeed built in opposition to public broadcasters, seen as in vassalage to governments. It is, however, crucial to insist on the *perceived* image, as the reality of relationships between radio stations, public or not, and governments, was a much more complex one. Nevertheless, this distinction found its roots, once again, in the geopolitical situation, as, due to their localisation, the stations avoided state monopoly.

Part of the popularity of commercial radio stations, especially with a younger audience, could be linked to their attractiveness as exotic and different, in the way they sounded. Their position as *radios périphériques* allowed them to develop a different and less-highbrow radio culture, by playing on their foreignness and turning it into something positive. An example which furthers the understanding of foreignness as a positive and exotic trait for radio can be found in a specific programme for instance. Towards the end of the period studied in this work, Europe n°1 hosted a show by the French disc-jockey known as Hubert. During this pre-recorded programme,[13] the host made sure listeners knew he was in Tahiti,[14] mentioning the beauty of the island on numerous occasions, and

[10] The *Office de Radiodiffusion-Télévision Française*, known as the ORTF, was the French public institution in charge of radio and television. It replaced the *Radiodiffusion-Télévision Française*, or RTF, in 1964, and existed until 1974. The acronym (O)RTF is used, in this book, to refer to both institutions.

[11] Richard Legay, "The Role of Commercial Radio Stations in the Media Vacuum of Mai 68 in Paris", *VIEW Journal of European Television History & Culture* (6:12, 2017).

[12] Although they had to maintain a certain neutrality and self-censorship, for they were 'tolerated' by the French government.

[13] *Hubert*, 1970, Europe n°1, 00:31:34, Archives of Europe 1, Paris. The programme presented by Hubert is analysed, in length, in the next chapter on the soundscape of commercial radio stations.

[14] Tahiti is an island located in French Polynesia, in the Pacific Ocean. Famous for its beauty, the island was annexed as a French colony, and became an overseas territory in 1946. Strictly speaking, Tahiti was not, in the 1960s, a foreign place for French listeners, as it was part of the same country. However, the island had (and still has) a strong exotic image in the French imagination, making it relevant in a section looking into foreignness.

insisting on his relaxed lifestyle. By doing so, he cultivated an exotic and positive image of his programme, in order to make it more attractive.

Hilliard and Keith find localism in radio when the medium broadcasts programmes which include features such as 'local music groups, panel and quiz shows with local people, talks and public affairs programs'.[15] Many programmes broadcast on commercial radio stations included some of these features, but with a major difference. Shows with local interests were often changing location, particularly in France, where they sometimes took an itinerant tradition. Among others, Radio Luxembourg's *Radio Circus* is a perfect example of an itinerant programme. First broadcast in 1947, the show was recorded under a circus tent, with a live audience, turning 'listeners into viewers'.[16] The station worked hand in hand with a performing troupe to broadcast a circus, variety, and game show on its waves.[17]

The popular music programme *Balzac 10-10* also included, for a while, a touring element, during which the show was recorded in small villages around France. Local groups and a live audience from the area were key features of the programme, anchoring it into a tradition of local radio. However, one show appears particularly insightful when it comes to localism. Europe n°1's *Bonjour Monsieur le Maire*, presented by Pierre Bonte,[18] is indeed a striking example of the station embracing localism as a key feature. In this programme, the host presents a different

[15] Robert Hilliard & Michael Keith, *The Quited Voice: the Rise and Demise of Localism in American Radio* (Carbondale: Southern Illinois Press, 2005), p. 56.

[16] Andreas Fickers & Pascal Griset, *Communicating Europe. Technologies, Information, Events* (Basingtoke: Palgrave Macmillan, 2019), p. 214.

[17] The history of Radio Circus and game shows on French radio was studied by Marie-Paule Schmitt. M.-P. Schmitt, "Les Jeux sur les ondes de la Libération aux années soixante-dix: grandeur et décadence d'un genre radiophonique", *Sociétés & Représentations* (24:2, 2007), pp. 353–369 & M.-P. Schmitt, "Les Jeux radiophoniques, en France, dans les années soixante", *Sociétés & Représentations* (37:1, 2014), pp. 219–228.

[18] Pierre Bonte (born in 1932) is a French journalist, who started in the regional newspaper *Ouest-France*, before joining Europe n°1 in 1959, where he presented *Bonjour Monsieur le Maire* for over a decade. He later worked for various French television channels. In 2019, he was invited to a commemoration at Europe n°1's former transmitter site in Saarland to celebrate his work on *Bonjour Monsieur le Maire*.

commune[19] in France each time, and some of its inhabitants are interviewed, for roughly ten minutes. The programme started in 1959 and lasted roughly 15 years, with a total of around 4,000 *communes* presented on air, making it an institution of the station throughout the Long Sixties. By systematically reflecting the 'needs, interests and culture of the local community [...] and featuring local people',[20] *Bonjour Monsieur le Maire* definitely belongs to local radio, even though the location appears to have been changing for each issue of the programme. Most of the issues of the programmes appear to share similar traits and follow a similar pattern, which might explain the long lifespan of the show, as all the *communes* presented shared commonalities and differences. They were unique in the sense that it was always a different *commune*, however, the issues mentioned would often be similar between the locations, therefore creating a sense of familiarity alongside a systematic renewal.

Among the 4,000 possible episodes of the radio programme, one stands out as it is focused on a small village—Rémering-lès-Hargarten—not far from Metz and especially close to Europe n°1's transmitter in Germany, making it an interesting illustration of the tensions between local and foreign. In this programme, the host followed a well-known pattern and interviewed a young 23-year-old man, keen on local history, who presented the ancient roots of his village. Meanwhile, the mayor, interviewed afterwards, shared more contemporary anecdotes, particularly on World War II and how 40 young men of the village—himself included—were integrated into the Wehrmacht, sent to the Russian front, and managed to hide in the local woods after their return to the village from their first leave. The mayor also mentioned the current growing difficulties of local young men as mines were closing, connecting it to problems affecting the entire region of Lorraine. These anecdotes were fully anchored into local history and the insecurities of the village, while the strong Moselle accent of the mayor is another indicator of the localism of the programme's 'sound of local accents'.[21] On the other hand, the

[19] The French word *commune* has to be understood here in the sense of a municipality, and can either be a small village or a town.

[20] Chignell, *Key Concepts in Radio Studies*, p. 132.

[21] Chignell, *Key Concepts in Radio Studies*, p. 132.

introduction of the show by the host is a clear reminder of the transnational potential of the area, as well as the transnational origins of Europe n°1:

> 'Pierre Bonte, reporting. For us, at Europe n°1, the mayor of Rémering-lès-Hargarten, is, in a way, like a neighbour. Rémering-lès-Hargarten is indeed the French municipality that is the closest to our Felsberg transmitter. Our transmitter, as you know, is located in Germany, simply because commercial stations are not allowed to install transmitters on French territory, and it is located right by the Franco-German border. And Rémering is on the other side of this border, on the French side, at exactly 40 kilometres in the North-East of Metz, in the Moselle region.'[22]

It is important to mention that references to transnationalism such as this one are rare in a programme like *Bonjour le Maire*, as it is usually solely focused on catering to local needs. Hilliard and Keith's definition of localism was grounded in the United States, where the number of radio stations was particularly high, meaning stations could cover local expectations.[23] In comparison, however, the European situation, with fewer stations, made this impossible, although commercial stations were also driven by the same interest in responding to the audience's taste. This difference between the two regions could explain why commercial radio stations developed this 'localism on the move'. As they could not develop a network of local stations, they responded to the wish for localism through travelling shows around a vast territory, in order to cater to expectations at a local level.

The 208 Exception. A key difference between the French and English services of Radio Luxembourg in the 1960s is that the 'station of the stars' had most of its programmes made in their studio located in Luxembourg, at the Villa Louvigny, and not in Britain, where most of the audience was. As mentioned previously in this chapter, the station also had offices in London, on Hertford Street, and this complex duality between Luxembourg and Britain was present even in the shows. References to both Britain and Luxembourg are common among the surviving audio sources, taking part in the shaping of a dual geographical attachment of

[22] Bonjour Monsieur le Maire (Rémering-lès-Hargarten), Europe n°1, 1965 or 1966, 00:10:01, personal archives of Mr. Roland Biesen.

[23] Hilliard & Keith, *The Quited Voice*, p. 56.

the station. The studio at the Villa Louvigny was regularly mentioned, in order to root the station as foreign and well established in the Great Duchy. For instance, in August 1965, Barry Alldis, presenting a 'Top 20' programme, opens his show in the following way: 'And this is your D.J., B.A., Barry Alldis, who welcomes you to the show [...] coming direct [sic.] your way from our studio, here, in the heart of Luxembourg'.[24] Anchoring a radio programme into the solid everyday world is common practice in the medium, as a way to counterbalance its blindness. It is also a way to indicate that 'broadcasters are not just "voices in the ether" but people like us'.[25]

This curated image of a station strongly rooted in Luxembourg is fascinating, especially because it was complemented by British references. In the same programme, the host advertised a variety of products, including the British brand of shampoo *Sunsilk* and a bingo game based in Durham, revealing that the targeted audience was indeed in Britain. It also revealed that flows of capital, through advertising, also originated from Britain. Other references to Britain were found in numerous programmes, with various forms. For instance, in *Top 30* in 1971,[26] Mark Wesley explained he was taking over another disc-jockey, Dave Christian, who was on holiday, and invited listeners to look for him if they were in the area of Worthing, in Sussex. Later in the same programme, listeners were invited to take part in a guessing game and had to send postcards with answers to an address in London.[27]

The news broadcasted on Radio 208 was supplied, from 1968, by the *Daily Mirror*, a major British newspaper.[28] It is difficult, however, to define these examples as belonging to radio localism, as they did not focus on the local level, but rather a national, and therefore do not fit so far with definitions used by Chignell or Hilliard and Keith. They did, however, anchor Radio Luxembourg into the British media landscape. This created

[24] *Top 20*, presented by Barry Alldis, 1st August 1965, Radio Luxembourg 208, 00:56:18, personal archives of Marius Zuiraitis.

[25] Crisell, *Understanding Radio*, p. 7.

[26] *Top 30*, presented by Mark Wesley, 28th August 1971, Radio Luxembourg 208, 00:28:49, personal archives of Marius Zuiraitis.

[27] More precisely, the address is London W1Y 8BA, which corresponds to Hertford Street, where offices of Radio Luxembourg were located.

[28] Richard Nichols, *Radio Luxembourg: The Station of the Stars* (London: W.H. Allen, 1983), p. 155.

a carefully crafted image of the 'station of the stars' as oscillating between British references and a materiality strongly rooted in Luxembourg. It usually followed a regular pattern: the host talked, *from* Luxembourg, *to* an audience in Britain, *thanks to* British sponsors. These grammatical choices reflected, therefore, the reality of the English service of the station, whose transnational connections between the United Kingdom and Luxembourg were not hidden, but present in the programmes, and even constituted part of the station's identity on air.

As this first part of the chapter has shown, commercial radio stations displayed many features that could be associated with localism and foreignness, through their programmes. While, at first glance, these two concepts seem to be juxtaposed, they actually worked well together. In *Modernity at Large*, Arjun Appadurai develops the idea that locality can be used to disguise transnational flows.[29] As seen above, commercial stations did engage with programmes providing such a 'spectacle of local'—to use Appadurai's phrasing—sometimes rather literally, as in the examples of *Balzac 10-10* and *Radio Circus*. By doing so, they hide the transnational flows of airwaves and capital operating in the stations,[30] to focus instead on local needs. Interestingly, they did not do so systematically, as they also crafted an image of transnationalism and foreignness, which contradicts this idea of disguise. The image of the pendulum used in the introduction of this section is therefore an ideal tool to reconcile the two elements. The stations indeed used both images, but never at the same time, in the same programmes. Europe n°1's audience for *Bonjour Monsieur le Maire* was different from the one tuning in to listen to Hubert, they had different expectations, therefore one show would respond to local interests, while the other would display an exotic image. Commercial radio stations were then able to balance both images by going back and forth in their programmes, offering a rather singular take on the questions of localism and globalisation in radio in the wider media landscape of the studied period. The case of the English programme of Radio Luxembourg did not, however, fit with the analysis above, due to a minor physical presence in Britain. It was also due to it having less air time,

[29] Arjun Appadurai, *Modernity at Large: Cultural Dimensions of Globalization* (Minneapolis: University of Minneapolis Press, 2010), p. 42.

[30] David Dominguez Muller, *Radio-Luxembourg. Histoire d'un média privé d'envergure européenne* (Paris: L'Harmattan, 2007), p. 51.

and a focus on popular music, offering less diversity in its entertainment programme. The question of localism and foreignness echoes, by extension, other fundamental aspects of transnational history. The duality between space and place established by Appudurai is particularly relevant here, as 'space' refers to the global and transnational level, while 'place' is directly linked to the local, which is illustrated by the pendulum described in this section. Discussing sound and communication technologies, Michael Bull explains that 'they inform us about how users attempt to "inhabit" the spaces of the city they move through. Mobile privatisation is about the desire for proximity, for a mediated presence that shrinks space into something manageable and habitable. Sound, more than any other sense, appears to perform a largely utopian function in this desire for proximity and connectedness'.[31] Therefore, the pendulum oscillating between the two notions of localism and foreignness discussed here takes a stronger role, as it becomes a tool for listeners to appropriate the transnational broadcasting space by '[shrinking it] into something manageable and habitable'.[32]

2 Relationships with Offshore Stations and Circulation of Disc-jockeys

Offshore radio stations, such as Radio Caroline and Radio London, were non-negligible actors in the media landscape in the Sixties.[33] Though they mostly found their audience in the British Isles, their reputation also reached France, where the ORTF broadcast on television, in 1966, a 20-minute documentary about offshore stations, with a focus on Radio Caroline and Radio City, on French television.[34] In this documentary

[31] Michael Bull, "Sound, Proximity, and Distance in Western Experience", in Veit Erlmann (ed.), *Hearing Cultures. Essays on Sound, Listening and Modernity* (Oxford: Berg, 2004), pp. 173–190.

[32] idem.

[33] While offshore stations broadcasting to England seem to have a prominent place in memory, other stations such as Radio Veronica, broadcasting off the Dutch coasts, should also be mentioned. In Ireland, radio stations also broadcast across the border between the Republic and Ulster, mostly with a political purpose. See Eddie Bohan, *A Century of Irish Radio, 1900–2000* (self-pub., McCawley Layout & Print, 2018), pp. 522–540.

[34] Pop pirate, Zoom, 28th July 1966, 00:21:38, INA.fr [https://www.ina.fr/ina-eclaire-actu/video/cpf07011433/pop-pirate] (last consulted 23 February 2023).

both pirate stations are portrayed in a positive way, particularly Radio Caroline, with shots of the ship and of the disc-jockeys at work. A lengthy interview, in French, of one of them—Rosko[35]—is central. At first, he is in front of his turntables, then he is seen sitting in his cabin, half-naked, a beer bottle in his hand, with another disc-jockey. In both shots, the young American man talks about his job, and, especially, about the differences between being a disc-jockey and a radio host, as he sees himself as focusing on the music, with little speech, while French radio does, according to him, the opposite. Interestingly, he joined RTL, a French-speaking station, later that year to present the show *Minimax*. Ronan O'Rahilly,[36] the creator of Radio Caroline, is also interviewed in London, and he can be heard—behind the French dubbing—likening his station to Voice of America, which is an interesting comparison with another major actor of the media landscape.[37] In contrast, the documentary also includes an interview with Gerard Mansell,[38] the BBC Deputy Managing Director, who explains, in French, the BBC's position towards the pirate stations: they are fighting against stations which use frequencies that were not allocated to them, making them illegal. In many ways, the documentary shows the offshore station in a favourable light: its disc-jockeys appear friendly, dressed casually—when dressed at all—and playing popular records. Whereas the BBC is portrayed in a rather unflattering manner: the BBC man appears stern and highbrow, the camera angle is more traditional, and the extracts from the station are from a religious programme.

This documentary is helpful as it reveals an interest in France for offshore stations, but, mostly, it points to the connections and influences

[35] Michael Pasternak (born in 1942), known as either *Président* (in French) or (Emperor in English) Rosko, is a American disc-jockey, who became known when he joined Radio Caroline in 1964. He also worked for BBC Radio 1 and for RTL, where he hosted, in French this time, the popular *Minimax* programme.

[36] Ronan O'Rahilly (1940–2020) was an Irish businessman and producer, mostly known for founding Radio Caroline.

[37] This comparison is likely built on the idea that both Radio Caroline and Voice of America were broadcasting to countries from beyond their borders, putting back-to-back Western States broadcasting monopolies to socialist party-states.

[38] Gerard Mansell was identified thanks to the precious help of Tony Stoller and Hugh Chignell. Mansell was born in France, and was a regular listener of the ORTF, which seems to match the man in the video, who speaks fluent French. See Chignell, *Public Issue Radio*, p. 84.

within a wider media landscape. Part of these transnational circulations included offshore pirate radio stations, whose disc-jockeys sometimes worked for Radio Luxembourg in the 1960s. Among some of the famous pirate voices, who also went to work in the Grand Duchy, one could mention Tony Prince[39] and Keith Skues,[40] who both had shows on air for '208', the English service of Radio Luxembourg. The American disc-jockey known as 'Emperor Rosko', mentioned previously, became known to the French-speaking audience as '*Le Président* Rosko' when he joined RTL, with his show *Minimax*, from 1966 to 1968. This flow of people from offshore stations to Radio Luxembourg is another aspect of the transnational dimension of the commercial radio station. The disc-jockeys were not simply skilled staff, they were also radio celebrities, whose image and identifiable voice travelled with them. Moreover, the aura of offshore pirate radio hosts appears to have crossed multiple borders, revealing this transnational entwined ecology, described by Föllmer & Badenoch.[41]

The case of Emperor, or *Président*, Rosko, is in that regard particularly rich, as it crossed a language barrier, switching from radio programmes in English to *Minimax*, in which he spoke fluent French while cultivating a form of exoticism due to his accent.[42] This show, broadcast in the late afternoon, was designed as an answer, from RTL, to fight back the popularity of Europe n°1's key youth programme, *Salut les Copains*, meaning that *Minimax* was part of the spearhead of RTL's 1966 move to expand

[39] Tony Prince (born Thomas Whitehead in 1944), is a British disc-jockey, who worked on Radio Caroline in the 1960s, where he became known as "The Royal Ruler", before joining Radio Luxembourg after the Marines Offences Act of 1967. He eventually became programme director of the English service in the late 1970s.

[40] Keith Skues, MBE (born in 1939), is a British radio broadcaster who started his career in the British Forces Network, while stationed in Germany. He later joined Radio Caroline in 1964, then Radio Luxembourg and Radio London, and was one of the first presenters on BBC Radio 1 after its launch in 1967.

[41] Föllmer & Badenoch, *Transnationalizing Radio Research*, p. 12.

[42] With the exception of local and regional stations—which appeared after the period studied in this monograph - distinctive accents, whether regional or foreign, have always been rare on French radio at national level. See Nozha Smati & Anne-Caroline Fiévet, "A la radio, la voix donne à écouter et à voir", 04/12/2017, INAglobal.fr [http://www.inaglobal.fr/radio/article/la-radio-la-voix-donne-ecouter-et-voir-10010].

their audience.[43] He was put on air at a time of peak listening numbers and was provided with expensive equipment and a studio, showing he was seen as a crucial investment.[44] Therefore, it is clear the American disc-jockey's reputation was known beyond the English-speaking sphere. This supports the idea that there were not only transnational flows of hosts and airwaves but also a shared set of cultural references, across borders and languages, as shown by the international reputation of an individual like Rosko.

Based on archival sources, the directors of Radio Luxembourg seem to have been keeping an eye on the evolution of offshore stations, however, their perspective appears to have been rather neutral towards them. This was illustrated, for instance, by a simple acknowledgement of the new programmes in English of Radio Veronica,[45] which broadcast from 1960 to 1974,[46] or a mention of the Council of Europe's attempts to stop pirate station activities.[47] Radio Luxembourg saw these stations as minor competitors, compared to the way it perceived Europe n°1, towards which it concentrated its focus on rivalry and monitoring. Radio Veronica, broadcasting mainly in Dutch, was seen as a competitor for the Flemish listenership.[48] The closing of offshore stations outside the United Kingdom did not affect the decreasing profitability of the

[43] A previous attempt to challenge *Salut les Copains !* was made with *Balzac 10-10*, however, Rosko's *Minimax* was a more successful one. Anna Jehle, *Welle der Konsumgesellschaft. Radio Luxembourg in Frankreich, 1945–1975* (Göttingen: Wallstein, 2018), p. 243.

[44] Rosko was "poached" from Radio Caroline to RTL, where the station offered him the first automated studio in France, tore down walls to give him more space, and provided him with a chauffeur and a secretary, reinforcing his celebrity status, and the efforts of the station to recruit him. See Jean-Pierre Defrain & Jacques Boutelet, *RTL: 40 ans ensemble* (Paris: Calmann-Lévy, 2006), p. 23.

[45] Minutes of the "Comité de direction", 28th January 1961, RTL Group Archives, Luxembourg.

[46] Radio Veronica was an offshore radio station broadcasting mostly in Dutch off the coast of the Netherlands. It operated from 1960 until 1974.

[47] Minutes of the "Comité de direction", 19th May 1964, RTL Group Archives, Luxembourg.

[48] Minutes of the "Comité de direction", 29th January 1965, RTL Group Archives, Luxembourg.

English service, as was hoped, due to the closeness of the target audience.[49] One hypothesis that could be drawn from these examples is that offshore stations were perceived as a threat, based on legal and ideological values, by public broadcasters and institutions, but not so much by commercial stations.

3 The American Influence and the BBC's Close Monitoring

Hiring Rosko to host a show on RTL was not a fluke for the station, which turned, at times, towards the United States for inspiration. In 1971, Jean Prouvost asked one of his representatives living in New York to prepare a lengthy report on American radio, which would be used among the board of directors of the station to, hopefully, fight their difficulties as a radio station, and the drop in audience share the previous year.[50] The extensive report opens with a reminder that there was a strong difference between American and French radio: the absence of a monopoly in the United States, and, therefore, a greater number of stations. These stations were said to strongly focus on reaching specific target audiences, rather than trying to 'please everyone', even if it meant alienating other social groups. Programmes and advertisements were decided based on this choice, with a general rule: the younger the listenership, the lesser the amount of speech in order to play more music. Something which echoes what Rosko said in his interview for *Zoom*. On the other end of the spectrum were 'all news' stations, which focused specifically on information and speech, and, as a consequence, targeted listeners over 40 years old. Based on an analysis of similar radio stations in the same category, the author of the report draws a series of conclusions to explain their success. It relied on a few elements: (1) the quality of the material (understood

[49] Explanations found in the source suggest that it was due to the devaluation of the pound, which reined profitability in. Another suggested hypothesis was that British television broadcasted its best programmes at the same hours as the English programme. The board of directors seems to have neglected another possibility. Indeed, the closing of offshores stations roughly coincided with the launch of BBC Radio 1, where many pirate disc-jockeys worked. This could have very well drained listeners of Radio Caroline and other similar stations. Minutes of the "Comité de direction", 30th April 1968, RTL Group Archives, Luxembourg.

[50] Report on "La Radio américaine", in Minutes of the "Comité de direction", 16th February 1971, RTL Group Archives, Luxembourg.

here as the kind of records played); (2) the host's personality, whose on-air persona depended on the type of station (exuberant and fast for young listeners, slower and calmer for older ones, etc.), who should be seen as an investment, as the longer someone works with the station, the better he is recognised by listeners; (3) the promotion and advertisement for the station; and, lastly, the sonic personality of the station, which includes jingles, and other sonic icons (although, the term is not directly used). It is on this last point that the author insists. In his view, the 'sound of the station'[51] is not as developed in France as in the United States, where listeners can identify stations easily, due to the content—rather than solely on sonic icons. Among his examples are references to stations playing folk music, specialising in Top 40, and Black radio,[52] which all point at a deeper specialisation of radio stations.[53] One crucial remark made by the author, who quotes a discussion with an unnamed 'radio specialist', is that the earlier development of television in the United States in the 1950s diverted potential audiences from radio, which suffered until rock-'n'roll came along and saved it. Though suspicious about the findings, the author still finds it inspiring, and draws his own conclusion: radio is the most qualified medium (in opposition to press and television, which was developing in France at the time) to deliver a youth market (understood as teenagers and 18 to 30 years old) to advertisers. Other parallels are drawn between the two countries, to explain the potential of radio against television, such as the growing mobility of listening, thanks to transistors and car radios for example.[54]

[51] In English in the original source.

[52] As described by Art M. Blake, the development of Black radio in post-war America is one key example of how the United States "sounded different" after World War 2. The Black CB (Citizens' Band) culture knew, according to the author, a strong increase in the 1970s, something that would likely sound quite different for a listener used to European radio. See Art M. Blake, *Radio, Race, and Audible Difference in Post-1945 America. The Citizens Band* (Basingstoke: Palgrave Macmillan, 2019), p. 40.

[53] Radio Luxembourg's model of seeing its audience as a whole proved viable in the period, due to the limited number of broadcasters in France, therefore, there was no need to specialise as radio had to in the United States. Jehle, *Welle der Konsumgesellschaft*, p. 249.

[54] If listening to the radio in cars has been possible since the 1930s, this only became more common in the 1950s and 1960s. The development of car radio using the transistor technology played a strong role in the United States, however, not as much in France, where transistor set manufacturers adapted by offering products adapter to both the

Radio in the United States and France have a long history of influencing one another over the twentieth century. The work undertaken by Derek M. Vaillant highlights perfectly how a transnational, and transatlantic, dialogue operated between both countries, in what he called 'The International Age of Radio'.[55] Therefore, the fact that Radio Luxembourg turned to American Radio for inspiration is no surprise and is an expression of this dialogue. The fact that the station looked across the Atlantic for solutions to a problem in the French market particularly highlights the various scales at stake in this situation.

The case of the *British Broadcasting Corporation* and its monitoring of commercial radio stations is used in this section to highlight a foreign institution's perspective on the stations studied. The choice of the BBC was made as the Corporation had regular interactions with Europe n°1 and Radio Luxembourg. Moreover, the relationship was less intense than between the French public broadcaster and the *périphériques* for instance. According to files kept at the BBC Written Archives Centre, the Corporation followed the early days of Europe n°1 closely, and, throughout the Long Sixties, seems to have had regular interactions with the station and its representatives. The relationship was not one of equals, as the BBC often looked down on Europe n°1, which saw its request for favours being regularly turned down. In a note for policy guidance dated 1959, the guideline to be adopted by the BBC regarding Europe n°1 was to refuse their requests to use facilities; which were thought to be unlikely anyway due to the station's conflict with the EBU.[56] On the few occasions Europe n°1 asked the BBC for the possibility of using some of their facilities to cover sporting events,[57] the requests received negative answers following this guideline. Another event that occurred in August 1965 helped understand the BBC's perspective on Europe n°1. Two disc-jockeys from France, staying in London, asked the reception

automobile and the portable usage. Elvina Fesneau, *Le Poste à transistors à la conquête de la France* (Bry-sur-Marne: INA Editions, 2011), pp. 184–190.

55 Derek M. Vaillant, *Across the Waves: How the United States and France shaped the International Age of Radio* (Chicago: University of Illinois Press, 2017).

56 Policy Guidance on USSR and satellite countries, 24th February 1959, E1/1, 929/1, Germany/Europe No 1, BBC Written Archives Centre, United Kingdom. Despite the name of the file, the guidelines covered more countries than those of the Eastern bloc.

57 The rejections were in relation to football and rugby games happening in the United Kingdom, which Europe n°1 wanted to cover, therefore needing facilities on site.

of Broadcasting House if they could witness a broadcast of the Light service programme. Internal correspondence reveals the BBC decided that it would not be a problem to let them watch a BBC programme but it should be made clear to the BBC that they were from a commercial radio and not the ORTF, meaning that they should not receive special help.[58] This infers the existence of a hierarchy among radio stations, in which commercial radio did not hold a prominent spot.

These two examples above shape the reasoning behind the BBC's attitude of disdain towards Europe n°1, which was based on two main arguments. First, the tensions between Europe n°1 and the European Broadcasting Union,[59] an institution Europe n°1 was not a member of,[60] creating, from the BBC's point of view, an absence of protections in working with Europe n°1.[61] Second, the BBC chose to favour its relationship with France and the (O)RTF, which could be damaged by actively working with Europe n°1.[62] This rather negative vision of Europe n°1 was the norm throughout the period, however, the BBC also acknowledged the quality of the news produced by the commercial station[63] as well as their command of phone-ins during live broadcasts. Despite the anecdotal aspect of this interest in phone-ins, what is striking is the wish

[58] Internal note regarding Visiting Disc-Jockeys from France (Europe No.1), 17th August 1965, E1/1, 929/1, Germany/Europe No 1, BBC Written Archives Centre, United Kingdom.

[59] Jennifer Spohrer extensively wrote in her doctoral dissertation about the tensions, in the early days, between Radio Luxembourg and the International Broadcasting Union - the predecessor of the EBU - and the International Telecommunication Union. See Jennifer Spohrer, "Ruling the Airwaves: Radio Luxembourg and the Origins of European National Broadcasting, 1929–1950", PhD dissertation (Columbia University, 2008).

[60] Europe n°1 joined the EBU, but only in 1978.

[61] Record of a telephone conversation between the European Liaison and Mr. Warman on the subject of facilities for Europe No. 1, 6th September 1962, E1/1, 929/1, Germany/Europe No 1, BBC Written Archives Centre, United Kingdom.

[62] At least, in the mind of the BBC staff.

[63] More precisely, the BBC acknowledged the quality of their foreign correspondents, taking the one in London as an example. They also acknowledged their coverage of the Algerian war, which resulted in tensions between the French government and commercial radio stations, as well as the censorship operated by France on its public broadcaster. For more details, see René Duval, "Europe n°1 et le guerre d'Algérie", in M. De Bussière, C. Méadel & C. Ulmann-Mauriat (eds.), *Radios et télévisions au temps des "événements d'Algérie" 1954–1962* (Paris: L'Harmattan, 1999), pp. 87–93.

of the BBC to investigate Europe n°1's techniques without their knowledge, asking their representative to be discreet and without consulting them in an official capacity. The BBC did not want to look like they were asking a favour or seeking technical advice,[64] likely in order to preserve a relationship in which they had the upper hand. This was indeed the core of the relationship between the BBC and Europe n°1, in which the commercial station was the one asking for favours, which were denied by the Corporation, which chose to preserve its relationships with France and the EBU, reflecting the complex dynamics at stage in the wider radio landscape of the period.

It appears that the BBC held a different view towards Radio Luxembourg, depending on the service, whether it was the French or the English one. The French service of Radio Luxembourg appears to have had a cordial relationship with the BBC. The State broadcaster allowed them to use their facilities on occasions, seemingly for sporting events. Interestingly, Radio Luxembourg was said to appreciate the BBC support and would have happily returned the favour at some point,[65] indicating a more cordial relationship than with Europe n°1. When the Paris representative suggested using Radio Luxembourg's facilities around Paris to cover the *Tour de France*[66] instead of those of the ORTF, the response from Broadcasting House, however, was an embarrassed one. Described as tricky, the question was met, after some discreet inquiries, with a negative answer.[67] While the relationship can be described as more cordial than with Europe n°1,[68] Radio Luxembourg was nevertheless not considered, by the BBC, as important as the ORTF. In a similar way as with Europe

[64] Correspondence from Mr. Best to the Paris representative, 19th April 1968, E1/1, 929/1, Germany/Europe No 1, BBC Written Archives Centre, United Kingdom.

[65] Correspondence from George Booth to Anthony Powell, 5th April 1967, E1/2,211/1, Luxembourg/Radio Luxembourg, BBC Written Archives Centre, United Kingdom.

[66] The *Tour de France* is a highly popular bike race, occurring almost every summer since 1903.

[67] The exact phrasing being: "I must reluctantly advise you against inviting Radio Luxembourg to provide facilities instead of O.R.T.F.". Correspondence from Anthony Powell to George Booth, 13th April 1967, E1/2,211/1, Luxembourg/Radio Luxembourg, BBC Written Archives Centre, United Kingdom. "BBC copyright content reproduced courtesy of the British Broadcasting Corporation. All rights reserved."

[68] Possibly because Radio Luxembourg joined the EBU from the start in 1950, therefore giving it a more respectable position, as well as "safeguards in dealing with them" to reuse the expression used against Europe n°1.

n°1, the BBC's guideline was to favour the French public broadcaster over commercial stations,[69] therefore reaffirming their peripheral position in the wider European radio landscape. It is harder to find evidence of the exchanges between the English service of Radio Luxembourg and the BBC, although the latter seems to have been following with interest the surveys on listenership ordered by the former, as they offered significantly different results from the BBC's audience research.

The three examples—the offshore stations, the American influence and the relationship between the BBC and commercial radio stations—developed above are particularly relevant to illustrate the complexity of foreign dynamics and influences which crossed the space in which commercial stations operated. The nature of the interactions is also diverse. They can take the shape of a transnational network of radio hosts, circulating among the various radio stations. They can also purposely look at another country and its radio system for inspiration, and bring back to Europe and France some aspects which would be adapted. And, lastly, it can also be a foreign institution monitoring, from Britain, what continental stations are doing. This illustrates how much the stations were actively working with a wide range of other actors, reinforcing the importance of transnational interactions on many levels.

4 The Competition between Europe n°1 and Radio Luxembourg

As the sections looking into the listenership surveys and the maps from the transmitter site in the previous chapter demonstrated, both stations kept an eye on each other, on their programmes, shares of audience, and technical reach, as they had a crucial impact on the stream of revenue. Even though many other actors were involved in the wider media landscape, it can be argued that the main dynamics occurred between Europe n°1 and the French service of Radio Luxembourg. These dynamics were mostly competitive, and took many forms, whether it concerned commercial activities and partnerships, programmes, the evolution of the 'sound' of the station and their technical progress. The competition did not appear to be aggressive, as some collaborative attempts have to be

[69] Policy Guidance on USSR and satellite countries, 24th February 1959, E1/1, 929/1, Germany/Europe No 1, BBC Written Archives Centre, United Kingdom.

mentioned. What is striking, however, is how much the stations developed in relation to one another.[70] One source reveals that the surveillance of the stations was not limited to numbers and commercial activities, but that it also covered the sound of the stations. In January 1966, a staff member of Radio Luxembourg spent their day listening, simultaneously, to Europe n°1 and RTL, and taking notes comparing how they sounded.[71] The notes detail precisely what the stations broadcast this day, from news flashes to number of records played, by way of adverts. In addition to this descriptive aspect, the notes include comments judging—rather harshly—the performances of RTL, in comparison to its counterpart. Traces of this activity of simultaneous listening further the idea that competition between both stations was a strong factor in development. In many ways, Radio Luxembourg and Europe n°1 became the reference point for one another, all the way to the very sound of the station, its flow, its use of music, and projected images through its programmes.

The competition between the stations represents an important facet of their relationship, however, as *radios périphériques* they also faced a complicated relationship with governments, especially the French one, and with its public broadcaster: the (O)RTF. Interestingly, the pinnacle of the tensions appears to have been during the protests of May 1968 in France, when the intense news reporting activity of the commercial stations led to numerous threats from the French government to end their right to broadcast to France.[72] It was during the same month that the stations broadcast a common show, during the night of the 8th May 1968, as a demonstration of their friendship. Sources demonstrate that negotiations had been ongoing for months before and after the common show, which was, in many ways, an audio materialisation of these talks. The fact that the programme was broadcast at the same time as important protest movements seems to be a coincidence.

[70] In this regard, the competition between the two *radios périphériques* also had an impact on the French public broadcaster, which had to renew and rethink its programmes in the 1960s to avoid seeing its audience moving to Europe n°1 and Radio Luxembourg. Jehle, *Welle der Konsumgesellschaft*, p. 361.

[71] "Résultats d'écoute comparés", 10th January 1966, RTL Group Archives, Luxembourg.

[72] Jean-Jacques Cheval, "Mai 68: un entre-deux dans l'histoire des médias et de la radio en France", *Comunicación* (2009: 26), pp. 13–25.

This chapter has shown, through diverse perspectives, how the transnational space of commercial radio stations was also the subject of external influences. By looking at the impact of other actors, such as the BBC, offshore stations and the wider American media landscape, this section reveals that the space studied is not hermetic. Moreover, external influences can induce change and clearly anchor commercial stations in a wider broadcasting space. This section also revealed the importance of competition between Radio Luxembourg and Europe n°1, as another key vector of change.

References

Appadurai, Arjun. *Modernity at Large: Cultural Dimensions of Globalization.* Minneapolis: University of Minneapolis Press, 2010.

Blake, Art M. *Radio, Race, and Audible Difference in Post-1945 America. The Citizens Band.* Basingstoke: Palgrave Macmillan, 2019.

Bohan, Eddie. *A Century of Irish Radio, 1900–2000.* Self-pub.: McCawley Layout & Print, 2018.

Bull, Michael. "Sound, Proximity, and Distance in Western Experience". In *Hearing Cultures. Essays on Sound, Listening and Modernity*, edited by Veit Erlmann, 173–190. Oxford: Berg, 2004.

Cheval, Jean-Jacques. "Mai 68: un entre-deux dans l'histoire des médias et de la radio en France". *Comunicación* 26 (2009): 13–25.

Chignell, Hugh. *Key Concepts in Radio Studies.* London: Sage, 2009.

Chignell, Hugh. *Public Issue Radio: Talks, News and Current Affairs in the Twentieth Century.* Basingtoke: Palgrave Macmillan, 2011.

Defrain, Jean-Pierre and Jacques Boutelet (eds.). *RTL. 40 ans ensemble.* Paris: Calmann-Lévy, 2006.

Dibbs, Martin. *Radio Fun and the BBC Variety Department, 1922–1967. Comedy and Popular Music on Air.* Basingstoke: Palgrave Macmillan, 2019.

Dominguez Muller, David. *Radio-Luxembourg: Histoire d'un média privé d'envergure européenne.* Paris: L'Harmattan, 2007.

Duval, René. "Europe n°1 et le guerre d'Algérie". In *Radios et télévisions au temps des "événements d'Algérie" 1954–1962*, edited by Michèle De Bussière, Cécile Méadel and Caroline Ulmann-Mauriat, 87-93. Paris: L'Harmattan, 1999.

Eck, Hélène. *La Guerre des ondes. Histoire des radios de langue française pendant la Deuxième guerre mondiale.* Paris: Armand Colin, 1985.

Fesneau, Elvina. *Le Transistor à la conquête de la France. La radio nomade (1954–1970).* Paris: INA Editions, 2014.

Fickers, Andreas and Pascal Griset (eds.). *Communicating Europe: Technologies, Information, Events*. Basingstoke: Palgrave Macmillan, 2019.

Föllmer, Golo and Alexander Badenoch (eds.). *Transnationalizing Radio Research: New Approaches to an Old Medium*. Bielefeld: Transcript, 2018.

Hendy, David. *Radio in the Global Age*. Cambridge: Polity Press, 2000.

Hilliard, Robert and Michael Keith, *The Quited Voice: the Rise and Demise of Localism in American Radio*. Carbondale: Southern Illinois Press, 2005.

Holmes, Su. *British Television and Film Culture in the 1950s*. Bristol: Intellect Books, 2005.

Jehle, Anna. *Welle der Konsumgesellschaft: Radio Luxembourg in Frankreich, 1945–1975*. Göttingen: Wallstein Verlag, 2018.

Legay, Richard. "The Role of Commercial Radio Stations in the Media Vacuum of Mai 68 in Paris". *VIEW Journal of European Television History & Culture* 6:12 (2017): 1–11.

Nichols, Richard. *Radio Luxembourg: The Station of the Stars*. London: W.H. Allen, 1983.

Peyrefitte, Alain. *Le Mal français*. Paris: Plon, 1976.

Schmitt, Marie-Paule. "Les Jeux sur les ondes de la Libération aux années soixante-dix : grandeur et décadence d'un genre radiophonique". *Sociétés & Représentations* 24:2 (2007): 353–369.

Schmitt, Marie-Paule. "Les Jeux radiophoniques, en France, dans les années soixante". *Sociétés & Représentations* 37:1 (2014): 219–228.

Smati, Nozha and Anne-Caroline Fiévet, "A la radio, la voix donne à écouter et à voir", 04/12/2017, INAglobal.fr [http://www.inaglobal.fr/radio/article/la-radio-la-voix-donne-ecouter-et-voir-10010]

Spohrer, Jennifer. "Ruling the Airwaves: Radio Luxembourg and the Origins of European National Broadcasting, 1929–1950". PhD diss., Columbia University, 2008.

Starkey, Guy. *Local Radio, Going Global*. Basingstoke: Palgrave Macmillan, 2015.

Vaillant, Derek W. *Across the waves: How the United States and France shaped the International Age of Radio*. Champaign: University of Illinois, 2017.

PART II

Sound

'This is the great miracle of wireless. The omnipresence of what people are singing or saying anywhere, the overleaping of frontiers, the conquest of spatial isolation, the importation of culture on the waves of the ether, the same fare for all, sound in silence'[1] —Rudolph Arnheim

This 'fare for all' is the main focus of the following chapters. 'Fare' is understood here as the sound coming out of the radio set, the transistor, what is heard and broadcast, the core of radio culture. Relying almost entirely on audio sources, the upcoming chapters give an overview of the key features of the various stations studied, identify what brought them together and differentiated them from public broadcasters as well as how they evolved. Other historians, such as Carolyn Birdsall in *Nazi Soundscapes*,[2] also used sound as a key material. However, they often did so by working with earwitnesses, and, here the focus is instead on a large number of sound archives. Therefore, the chosen approach looks more into what constituted the sound of commercial radio, rather than its impact. Due to the nature of some of the archival material—which was likely recorded by amateurs rather than by the stations themselves—the

[1] Rudolph Arnheim, *Radio*, translated by Margaret Ludwig & Herbert Read (London: Faber & Faber, 1936), p. 14.

[2] Carolyn Birdsall, *Nazi Soundscapes: sound, technology and urban spaces in Germany, 1933–1945* (Amsterdam: Amsterdam University Press, 2012).

analysis of the sound of commercial radio stations also includes non-diegetic[3] audio elements from the space of reception. These sounds (*e.g.* a chair being moved) are present in some of the recordings, and they are useful to provide information to researchers about the context of the listening experience at the time. These sonic elements act as a complement to the sound studied here, which primarily focuses on the programmes themselves, in all their sonic richness.

The analysis conducted in the following chapters finds many of its roots in sound studies. And, by doing so, it shifts the historian's perspective to the sense of hearing and to treat sound as a historical resource.[4] As mentioned by Goodale, there are several ways to study and analyse sound, hence why so many lenses and designations exist. Quoting the works of scholars such as Michele Hilmes, R. Murray Schafer, and Jonathan Sterne, Goodale explains that 'the viability of a specific method for reading sound is not as important as the greater argument that sound can be read'.[5] Reading sound is the precise intention here. A way to conduct such a reading of sound can be by treating historical recordings as 'radio texts', in order to 'seek an understanding of the aesthetic and formal qualities at stage in radio and relation to meanings, in particular cultural forms'.[6] By doing so, it is possible to work on both the form (the type of programme, the sound design, etc.) and the content (what is said, which songs are played, etc.) of the recordings used. Fortunately, this approach is supported by other disciplines, which the sound historian can draw from. For instance, radio studies have always underlined the importance of the voice, notably for its power over imagination.[7] While

[3] Michel Chion defines a non-diegetic sound as a sound "whose apparent source is not the space-time of the scene depicted". By opposition, a diegetic sound has an apparent source in the programme. Michel Chion, "100 concepts pour penser et décrire le cinéma sonore", bilingual version, October 2012. [http://michelchion.com/] (last accessed 27/03/2023).

[4] In opposition to the usual predominance of sight and what Greg Goodale coined 'the shackles of visual evidence'. Greg Goodale, *Sonic Persuasion: Reading Sound in the Recorded Age* (Urbana: University of Illinois Press, 2011), p. 5.

[5] *ibid.*, p. 12.

[6] Tim Crook, *The Sound Handbook* (Abingdon: Routledge, 2012), pp. 151–152.

[7] Nozha Smati & Anne-Caroline Fiévet, "A la radio, la voix donne à écouter et à voir", 04/12/2017, INAglobal.fr, [http://www.inaglobal.fr/radio/article/la-radio-la-voix-donne-ecouter-et-voir-10010].

anthropologists have made it the centre of their research,[8] scholars of film studies like Michel Chion have also offered insightful publications on the use of sound.[9] It is indeed crucial to recognise the power of sound, and not to neglect 'how aural phenomena and sensory experience in general may be historically constructed'.[10] Moreover, engaging in the study of sound requires a shift in the historian's perspective, which, according to Veit Erlmann, might bring 'new and richer kinds of ethnographic data', and 'force us to rethink a broad range of theoretical and methodological issues'.[11] Among such issues is the importance of technology, which cannot be 'bypass[ed]'[12] and is even more relevant and crucial when working on radio history.[13] Another methodological question pertains to the nature of radio. As mentioned by Cohen, Coyle and Lewty in their introduction to *Broadcasting Modernism*, radio is an 'especially ephemeral medium, incapable of inscription',[14] which makes its historical study particularly challenging. Fortunately, the case of commercial radio stations such as Radio Luxembourg and Europe n°1 in the Long Sixties provides the historian with a comfortable amount of sound recordings. Although these recordings may not inform researchers on the entire aural and cultural experience of radio, they nevertheless provide insights as to what was offered by commercial radio stations.

[8] See for instance David Le Breton, *Éclats de voix. Une Anthropologie des voix* (Paris: Métailié, 2011).

[9] See for instance Michel Chion, *Le Son: Ouïr, écouter, observer* (Paris: Armand Colin, 3rd edition 2018, 2010, 1998).

[10] David Suisman, "Introduction: Thinking Historically about Sound and Sense", in D. Suismand & S. Strasser (eds.), *Sound in the Age of Mechanical Reproduction* (Philadelphia: University of Pennsylvania Press, 2010), pp. 1–12.

[11] Veit Erlmann, "But What of the Ethnographic Ear?", in Veit Erlmann (ed.), *Hearing Cultures. Essays on Sound, Listening and Modernity* (Oxford: Berg, 2004), pp. 1–20.

[12] idem.

[13] Notably because sound technologies support the idea of the senses being interrelated, as suggested by Steven Connor, who reminds his readers that: "The senses are multiply related; we rarely if ever apprehend the world through one sense alone [...] the evidence of sight often acts to interpret, fix, limit, and complete the evidence of sound [...] the relations between sound and touch, by contrast, tend to be mimetic: touch accompanies, mimics, performs sound rather than translating or defining it". See Steven Connor, "Edison's Teeth: Touching Hearing", in Erlmann, *Hearing Cultures*, pp. 153–172.

[14] D. R. Cohen, M. Coyle & J. Lewty (eds.), *Broadcasting Modernism* (Gainesville: University Press of Florida, 2009), pp. 1–7.

CHAPTER 5

What Was the 'Sound' of Commercial Radio in the Long Sixties?

Commercial radio stations were known to sound different from public broadcasters, notably through their attention to light entertainment programmes and advertisements. This chapter builds on an in-depth analysis of a sample of historical recordings, taken from the wider corpus consulted in preparation for this book, in order to describe what these stations 'sounded like', and how this specific sound evolved throughout the Long Sixties. Nine recordings were chosen. Three per station: the English service of Radio Luxembourg, the French service of Radio Luxembourg, known as RTL from 1966, and Europe n°1. And three per period of the Long Sixties: the Early Sixties (c. 1958 to 1964), the High Sixties (1964 to 1968), and the Late Sixties (1968 to c. 1974), as developed in the introduction. The documents were chosen as part of this sample for multiple reasons. First, they are representative of light entertainment for each station and each period, and they were programmes often on air at popular slot times. News programmes are not among the sources. Instead, the chosen sources are entertainment programmes, such as record shows, programmes about musical performances, etc. Second, the sources were chosen for their sound quality and sufficient duration. This facilitates the process of a textual description, which offers a basis for a deeper analysis of the sample.

The clear limits of this sample of nine recordings obviously do not allow for an exhaustive description of everything that was broadcast on

R. Legay, *The Cultural Histories of Radio Luxembourg and Europe n°1*,
https://doi.org/10.1007/978-3-031-46250-4_5

the studied radio stations during the chosen period.[1] However, an analysis of the sample of the nine recordings reveals the elements that tie the different stations together under the umbrella of commercial radio. It becomes apparent the studied commercial stations share a common cultural fabric, with similar norms and practices that this sample can highlight. It also highlights the nuances and differences (other than the languages used) between the stations. Furthermore, the sample shows evolutions of 'doing' commercial radio throughout the period.

1 The 'Early Sixties'

The first period of the Long Sixties studied in this chapter was coined the 'Early Sixties', the period covering the years between 1958 and 1964. During these years, the heritage of the 1950s is still present in the sound of radio, however, the arrival of Europe n°1 in 1955 has already started to jostle the media landscape. Due to the nature of radio recordings, two of the three sources analysed below are from the end of the 'Early Sixties', as it was difficult to find recordings whose length and quality were sufficient enough for a detailed textual description. The third source is from 1957, however, the show (*Pour ceux qui aiment le jazz*) had a strong impact on popular culture and lasted for many years; therefore, it seemed to be an appropriate consensus to add it to this analysis of the sound of the 'Early Sixties'.

Pour ceux qui aiment le jazz, 1957, Europe n°1. The first recording is a programme in French about jazz music[2]—*Pour ceux qui aiment le jazz*—broadcast on Europe n°1 in March 1957.[3] The show first aired in

[1] According to the archives of RTL Group, the French service of Radio Luxembourg broadcast, on average, for 7900 hours per year between the years 1964 and 1971. There is no information regarding the year 1965 however. Minutes of the "Assemblée générale annuelle", 18th May 1965, 16th May 1967, 21st May 1968, 20th May 1969, 18th May 1971 and16th May 1972, RTL Group Archives, Luxembourg.

[2] Listening to jazz music on the radio was not uncommon for Western youth in the 1950s, as shown by Hilgert, who focused on Western Germany and the United Kingdom. Christoph Hilgert, *Die unerhörte Generation: Jugend im westdeutschen und britischen Hörfunk, 1945–1963* (Göttingen: Wallstein, 2015), pp. 134–136.

[3] Pour ceux qui aiment le jazz, 29th March 1957, Europe n°1, 00:19:17, Archives of Europe 1, Paris.

1955 and was presented, each day, by Daniel Filipacchi and Frank Ténot.[4] This analysis covers the first nineteen minutes of the programme,[5] in which the hosts, specialists in jazz music, play records and discuss them with each other. The recording begins on the jazzy opening title of *Pour ceux qui aiment le jazz*, which lasts slightly under a minute, before moving on to the voice of Frank Ténot asking his co-host: 'Tell me, Daniel, do you remember the year 1951, in France?' as a way to introduce a discussion on Duke Ellington, his concerts in the country that year, the negative reactions of the French opinion at the time, and the hosts' personal memories surrounding Duke Ellington in France. Ténot and Filipacchi then proceed to play one of the songs that Frank Ténot remembers being performed by Duke Ellington during one of his concerts. The score is a calm instrumental jazz tune,[6] followed by one of the hosts commenting to the other one: 'Tonight, you[7] see, we are opening with an atmosphere that is very... Very relaxed'. The two hosts then carry on discussing the discography of Ellington, who is referred to as *Le Duke* in their exchange.

The hosts describe the records they play (such as 'New York City Blues' and 'Clothed Woman'), both the song itself and its context, such as anecdotes on the recording conditions. Ténot and Filipacchi converse together as if they were having a private conversation about music. There are no references or addresses to listeners. Both men speak calmly, with clear pauses between records and speech, and the recording gives the impression of catching an intimate discussion between experts. This intimacy is due to the way they speak, but also to the way they address each other, by using their first names and the 2nd-person singular pronoun *tu*, signs of familiarity in the French language. Their positions

[4] Daniel Filipacchi (born in 1928) is a French editor and radio personality. Regarding his career with magazines, it should be noted that he worked as a photographer for the *Paris Match* magazine, then became editor of *Jazz Magazine* in 1956, with Frank Ténot (1925–2004), with whom he regularly collaborated. Filipacchi launched the *Salut les Copains* magazine in 1962, and acquired or launched a few other magazines throughout the 1960s, including *Mademoiselle Âge Tendre*, *Lui*, etc. He later bought *Paris Match* in 1974. Filipacchi and Ténot presented *Pour ceux qui aiment le jazz* on Europe n°1 from 1955 and *Salut les Copains* from 1959 to 1969.

[5] The source is only a partial recording of the programme, and stops after nineteen minutes.

[6] Unfortunately, it was not possible to identify it.

[7] In French, the man is using the pronoun *tu*, therefore addressing the other host.

as experts are shaped by the sheer amount of musical knowledge they display throughout the programme. This knowledge alternates between contextualisation of specific records—*i.e.* comments on the recording conditions—and connections with the artist's discography and with the wider jazz culture. Additionally, towards the end of the recording, one of the hosts goes into a lengthy explanation about the concept of walking bass,[8] reinforcing the perceived expertise.

Balzac 10-10, 1963, Radio Luxembourg (French Service). The second audio source is a mix of three different recordings of *Balzac 10-10*, a radio programme centred on popular music directed towards a youth audience.[9] Radio Luxembourg designed the programme, which aired from September 1963, as a way to reach a younger audience, and directly competed with Europe n°1's *Salut les Copains*.[10] According to Jacques Garnier, the creator of *Balzac 10-10*, this was done by turning towards the practices of American disc-jockeys and current popular music stars.[11] Usually, the show was recorded in Radio Luxembourg's studios in Paris, but, occasionally, the show would tour France and be recorded in its marquee in various towns, in front of a live audience. This was another strategy to compete with *Salut les Copains* and its 'summer podium' outside the studios.[12] In the archive consulted, the three extracts all came from shows which were recorded in 1963[13] outside of the

[8] A walking bass is a style of bass accompaniment reminding the rhythm of someone walking, quite popular with various musical genres, including jazz.

[9] Balzac 10-10, 1963, Radio Luxembourg, 00:50:24, INAthèque, Paris.

[10] With relative success however. See Anna Jehle, *Welle der Konsumgesellschaft: Radio Luxembourg in Frankreich 1045–1975* (Göttingen: Wallstein, 2017), p. 239.

[11] Denis Maréchal, *RTL. Histoire d'une radio populaire. De Radio Luxembourg à RTL.fr* (Paris: Nouveau Monde Éditions, 2010), p. 429.

[12] Idem.

[13] According to the metadata of the INAthèque, this recording is from 1963, however, Denis Maréchal, on the other hand, states, in his monograph on the history of the station, that *Balzac 10-10* replaced another programme, *Le passe-temps des dames et des demoiselles*, in 1964 (Maréchal, *RTL. Histoire d'une radio populaire*, p. 429). Robert Prot states another date, 1966, for this replacement. See Robert Prot, *Précis d'Histoire de la Radio et de la Télévision* (Paris: L'Harmattan, 2007), p. 171. Due to the conflicting information, it was chosen to trust the metadata on the recording.

Parisian studios, in smaller towns in France.[14] It appears that the recordings were made externally, by someone listening to the station at the same time, which could also explain why a single document is actually made of extracts of three different shows of *Balzac 10-10*. The metadata offers little information, and it is unclear why these extracts were mixed into one recording, and why these specific programmes were picked. There are no apparent connections between them; the locations are in three different places in France, the bands playing are different, etc. The extracts are all from musical talent shows, in which local amateur bands of young musicians perform in front of an audience. The three parts of the recording follow a similar pattern. The main host, Michel Cogoni,[15] introduces the programme and its concept, then moves on to introduce the various bands, who then play a few songs. There is a spectrum of voices heard throughout the recording, ranging from professional to amateur. The high level of professionalism is exemplified by the host's opening statement: 'Good evening everyone. Michel Cogoni wishes you a good evening in the company of the Fingers[16] [...] And the entire team, which is Jacques and I, the technicians, the secretaries; the whole team wishes you a... Very... Good... Evening [insists strongly on the sentence] with the Fingers'. Contrastingly, the microphone is regularly given to band members, who only say a few words to introduce themselves, the song they play, and the name of the band. They are far less used to public speaking, and the band members, mainly male teenagers, can barely be heard, especially in comparison to Cogoni, who often has to repeat loudly what they say. This amateurism of the band echoes their musical performances, which are often off-key and not clearly audible on air. For instance, in the second extract, a voice[17] can be heard asking the band members to speak closer to the microphone. In the middle of the last song performed, during the third extract, the bass player sounds confused and another band member can be heard telling him: 'Faster,

[14] In St-Quentin, in Northern France, in Firminy, by Lyon, and in one unidentified town in the French Ardennes.

[15] Michel Cogoni (1936–1969) was a French journalist and radio personality, who worked on most of the French-speaking commercial radio stations, Europe n°1, Radio Luxembourg, as well as Radio Monte Carlo.

[16] The band is called Fingers in French as well, although the host pronounces it with a heavy French accent.

[17] Likely a technician's.

faster!', reinforcing this amusing amateur characteristic. These unplanned sonic elements contrast strongly with the professionalism of the radio host—and of the station overall—giving the source an interesting tension between professionalism and amateurism. In the three extracts, the public is omnipresent on air, clapping, cheering, and singing at times, and constitutes a key sonic feature of the recording, even if always in the background.

Your Record Show, 1963, Radio Luxembourg (English Service). The third source is an extract from a record show, in English, presented by Brian Matthew on Radio Luxembourg 208 in 1963.[18] Entitled *Your Record Show*, this programme follows a typical format of this genre on the station; a succession of records[19] introduced by the disc-jockey, interspersed with advertisements and sponsored messages. In this format, the personality of the disc-jockey is central to the show and is part of a wider celebrity culture,[20] to which Brian Matthew belongs.[21] The recording is of, roughly, the second half-hour of *Your Record Show* only; it starts in the middle of a song—'Rainin' in my Heart' by Slim Harpo—and stops when Brian Matthew signs off before the following show, by Kent Walton. Most of the comments announce the artist and the title of the record, always in a different fashion. For example, Matthew introduces 'Hear my song, Violetta' by Ray Adams in this manner: 'Now, if there is a young lady called Violetta tuned to 208 at this moment, then this [pause] is [pause] just for you', speaking particularly smoothly in the last words. The host often leaves a short pause between his comments at the beginning and the end of a record, which clearly separates the different elements of the programme. This applies to advertisements as well, which are not read by the host, but by other male voices, always at a

[18] Your Record Show, presented by Brian Matthew, 1963, Radio Luxembourg 208, 00:25:44, personal archives of Marius Zuiraitis.

[19] These records include, among others, 'La Paloma' by Tony Hatch, 'Granada' by Frank Sinatra, 'Who Put The Bomp (In The Bomp, Bomp, Bomp)' by The Viscounts and 'Nanette' by The Sherwoods.

[20] Hugh Chignell, *Key Concepts in Radio Studies* (London: Sage, 2009), p. 18.

[21] Apart from his time on the British service of Radio Luxembourg, Brian Matthew was also posted in post-war Hamburg as part of the British Forces Network, and joined the BBC in the mid-1950s, where he later worked on the Light Programme, and then on BBC Radio 1, after its launch in 1967. See 'Obituary: Brian Matthew', *BBC News*, 8th April 2017, [https://www.bbc.com/news/entertainment-arts-39513249] (last consulted 3 March 2023).

slow pace, carefully enunciated. There are, in the recording, two adverts for the sponsor of the programme, the Senior Service Cigarettes, and a well-known ad by Radio Luxembourg listeners in the Fifties and early Sixties: the 'Infra-Draw Method', a betting technique developed by a man named Horace Batchelor. Batchelor's voice, quite slow, old, and with little emphasis, is heard in the advert: 'Welcome to the Horace Batchelor programme. And this is Horace speaking to you personally. I am here to help you tonight on the pools. And I can assure you my famous Infra-Draw Method can, and will, get you dividends. I am the world's record holder for number of wins on the pools. And users of my method have already won over 8 million pounds. [...] I have appeared on television as Britain's star football pool expert. And the amazing Infra-Draw Method is the secret of my successes. Recent wins of mine were £74,586, £45,286 and £5,500. And you can win these huge dividends yourself, if you will only use my Infra-Draw Method. This is the 13th season with the method, which is a huge success. And the name of Horace Batchelor needs no introduction to you. Now, would you like to win even small dividends? Of say, three, four or five hundred pounds quickly? If so, do please write me at once. Simply send me your name and address and I will send you complete details of this fabulous method absolutely free of charge'. Another, younger and more energetic, voice follows: 'Write at once for free details of the amazing Infra-Draw Method to Horace Batchelor, Department One, Keynsham, Bristol. Here it is once again. Horace Batchelor, Department One, Keynsham, spelled K-E-Y-N-S-H-A-M, Keynsham, Bristol'. This unusual and particularly lengthy advert is often mentioned in grey literature[22] and became famous[23] and iconic for the station to the point that it was referenced in a song.[24] The end of the recording concurs with the end of the programme, and one can hear Brian Matthew directly addressing the listeners: 'Thank you for your

[22] "This is the radio station which gave you the Ovaltinies, the *Top Twenty Show*, the Beatles, Horace Batchelor and the Infra-Draw method, [...]". Richard Nichols, *Radio Luxembourg: The Station of the Stars* (London: W.H. Allen, 1983), p. 8.

[23] Hence why it was decided to transcribe the advert almost entirely.

[24] The Bonzo Dog Doo-Dah Band made an album called Keynsham, in reference to the town of Horace Batchelor, whose constant repetition of the spelling helped put on the map, and, in this album, the first track, 'You Done My Brain In', starts with: 'I have personally won over...', a reference to the Infra-Draw Method radio advertisement. These details, though small, show how strongly this advertisement was engraved in a wider popular culture.

company [...] I shall look forward to our next meeting very much indeed [...] this is your old mate singing off', creating a sense of familiarity and proximity with the audience.

The three recordings analysed here represent a gateway into the sound of commercial radio in the 'Early Sixties' as they offer elements rooted in the longer heritage of radio, such as the slow, lengthy discussion about jazz between experts like Filipacchi and Ténot. Other aspects, *i.e.* the focus on youth audience in *Balzac 10-10* or the omnipresence of records as central objects in Brian Matthew's show, point at aspects of radio strongly rooted in the Long Sixties, and traits of the stations throughout the whole period.

2 The 'High Sixties'

The second set of audio documents is built in a way to highlight the 'High Sixties', from 1964 to 1968, a period of strong changes, notably in radio history. Radio Caroline, and other offshore stations, started to broadcast around 1964 and 1965, before having to shut due to the Marines Offences Act of 1967. In 1966, the French service of Radio Luxembourg became RTL and underwent many changes in its programmes and image, while the BBC launched BBC Radio 1 in 1967 in order to attract younger audiences. All of these elements together illustrate key changes in the radio landscape of the Sixties and the importance of the period between 1964 and 1968.

Salut les Copains, 1965, Europe n°1. The first recording is from one of Europe n°1's most famous youth programmes: *Salut les Copains*. The programme was broadcast on the 21st May 1965.[25] It is a rather long recording; its time is slightly under two hours. In opposition to most of the material in the archives on Europe 1, which was recorded by the broadcaster, this source was likely recorded from a listener's perspective. Recorded sources from the listener's perspective can be more intriguing because of the inclusion of elements that are usually omitted from broadcaster recordings. In this case, the recording includes the news before, and during, the show, as well as adverts interspersed throughout the programme. *Salut les Copains* was a radio programme targeting a young audience and centred around popular music. It was broadcast daily during

[25] Salut les Copains, 21st May 1965, Europe n°1, 01:45:55, Archives of Europe 1, Paris.

weekdays on Europe n°1 from 19th October 1959[26] to 1969, from 5 to 7 pm; in other words, a popular time slot for more or less a decade. The time slot partly explains how the show acquired its iconic status. The show's popularity[27] also led the producers of the show to publish an eponymous magazine, making it relevant to include one recording of *Salut les Copains*, to highlight one of the most listened to shows of that period. In this specific recording, the first several minutes provide a comprehensive overview of the entire programme. The listener can first hear the '*journal parlé*' of Europe n°1, the news, which is quickly followed by an introductory song of *Salut les Copains*,[28] a record ('Il te faudra chercher' by Richard Anthony), and, finally, the voice of the host, Daniel Filipacchi. He explains that the record he just played was the one he had issues with the day before, which is why he is playing it again, interestingly connecting the shows over different days and therefore referring to radio's temporal rhythms. Furthermore, he explains that this record is ranked number 10 in the *Salut les Copains*'s hit-parade, a reference to the eponymous magazine which published these hit-parades. Filipacchi introduces another record, played immediately: 'Dans mes bras, oublie ta peine' by Michèle Torr, which is followed by an advert for a stain remover product, read by a female voice, before a jingle is played, in which a chorus of electronically distorted young voices reply to each other to give the host's address.[29] Interestingly, the host adds a few comments to this pre-recorded jingle: 'This is the address you should always write to me, if you want to request a record that will be played on *Salut les Copains*, or if you want to talk about the magazine *Salut les Copains*, or about the magazine *Mademoiselle Âge Tendre*. By the way, let me tell you, we are in the period in which the new issue of *Mademoiselle Âge Tendre* is coming out. Incidentally, I have received some

[26] Prot, *Précis d'Histoire de la Radio*, p. 146.

[27] Chris Tinker, 'Shaping Youth in Salut les Copains', *Modern & Contemporary France* (15:3, 2007), pp. 293–308. The show is even called '*émission-vedette*', or 'star-show'. See Robert Prot, 'Salut les Copains', in *Dictionnaire de la radio* (Grenoble: Presses Universitaires de Grenoble, 1999).

[28] Some sort of extended jingle, a sonic icon marking the beginning of the programme.

[29] The mail addressed to Filipacchi is actually just sent to 'Daniel', potentially in order to make him more approachable, closer to his usually young audience. The address given in this jingle is the 'Postal Box 150' in the 8th *arrondissement* in Paris, and can be heard in other shows as well.

letters about this issue [...]'. He continues talking about the upcoming results of the *Mademoiselle Âge Tendre*'s election, a beauty pageant sponsored by the magazine, also published by Filipacchi. These comments are particularly interesting in regard to the dynamics operating around the host's various press activities, as he mentions magazines he owns during the programme. Moreover, the address to contact him appears, according to his comments, to be the same for the radio show and the two magazines, centralising letters from readerships and listenership into a single place. Coupled with the culture of celebrity around disc-jockeys and radio presenters, these comments illustrate a form of the ruling figure of 'Daniel' over a transmedial empire[30] and a large public. The rest of the programme is a succession of records,[31] some being ranked in the *Salut les Copains* hit-parade, which are interspersed with adverts and comments from the host. Roughly midway through the programme, Filipacchi mentions the *Mademoiselle Âge Tendre* magazine again, giving the names of the models posing in the most recent issue, including the French singer Sheila. Then, at one hour and twenty minutes into the show, the listener is invited to listen to an interview with two members of *The Shirelles*.[32] This interview is conducted by a man called Bernard from radio studios in New York, who is translating the answers into French, while his questions in English appear to have been cut.[33] The conversation between Bernard and *The Shirelles* revolves mostly around France, as the presenter asks about their favourite singers, the band's French tour, their memories of Paris, and their opinion on French rock. Interestingly, Bernard claims the interview is live, yet obvious edits of the discussion indicate it was not the case. The impression of liveness in radio is crucial to the medium, as, according to Chignell, it helps listeners to develop a sense of intimacy and co-presence.[34] It should be noted that a programme does not have to be live to give the impression of liveness, as there are

[30] Daniel Filipacchi published, apart from *Salut les Copains* and *Mademoiselle Âge Tendre*, other magazines such as the French version of *Playboy* and *Lui*, and owned a majority of shares of the *Cahiers du Cinéma* for a few years in the 1960s.

[31] Among others, listeners could hear a song by Sam Cooke and another by Sylvie Vartan.

[32] *The Shirelles* was an American girl band popular in the 1960s. Only two of its members took part in this interview: Doris Coley and Addie 'Micki' Harris.

[33] Based on the sound of the interview and the rhythm of conversation.

[34] Chignell, *Key Concepts in Radio Studies*, p. 88.

many ways to do so. However, such strategies might, as is the case here, blatantly fail. The last advert in the recording, which ends before the programme does, is about a competition for young people under sixteen, who have to draw what an umbrella in the year 2000 might look like. This is particularly interesting as the age of the listeners is often not mentioned, while this advert strongly targets a particularly young audience.

Disques sur 20, 1968, RTL. The second source is an extract of *Disques sur 20*, a French record show from 1968,[35] broadcast on RTL, and presented by Jean-Bernard Hebey, a well-known voice of light entertainment on commercial radio stations, who began his career on Europe n°1 in 1965, where he was involved on *Salut les Copains* for three years, before joining RTL in 1968 and becoming head of programmes in 1971.[36] This recording, part of the *Fonds Gaston L'Herbier*[37] at the INAthèque, includes non-diegetic sonic elements, the most apparent one being a constant humming, buzzing, and sound likely due to recording conditions at the time. The recording opens in the middle of the news, presented by a male journalist, who, in his closing remarks, tells listeners: 'I'm leaving you in the company of Jean-Bernard Hebey', to which Hebey immediately replies: 'Thank you, my good man!' before the jingle of the programme starts. The host introduces the programme over the jingle, and encourages listeners to request one of the two programmes for the show of the day. The first, *le programme anglais*, focuses on new records in English, and some French singers like Johnny Hallyday and Claude Nougaro, while the second, *le programme français*, includes, for example, the works of Georges Brassens and Jacques Brel. Listeners are encouraged—by the host—to call the station and request their favourite programme to listen to it for the duration of the show. In total, the host takes five callers, three women and two men, who all ask for the English-oriented programme. During the calls, instrumental songs are played in the background—including a contemporary cover of Ravel's 'Bolero'—demonstrating that there are no periods of silence during the programme,

[35] Extract from Disques sur 20, 23rd September 1968, RTL, 00:26:08, INAthèque, Paris.

[36] According to Hebey's personal webpage, [https://hebeypop.blogs.com/about.html] (last accessed 3 March 2023).

[37] Gaston L'Herbier was a technician who worked for RTL and made many personal recordings on the stations' programmes. His family later gave these recordings to the INAthèque.

not even between calls. This interactiveness in the selection of the records played during the programme is highly significant because it creates an unusual dynamic in which listeners appear to have some control over the content of the programme. The succession of records is interrupted on two occasions. The first time, Hebey host stops a record—'The House That Jack Built' by Aretha Franklin—rather abruptly, with some gurgling sounds followed by an announcement that the 'battleship game' is to be played next Wednesday, before playing a record by Johnny Hallyday. This abrupt interruption of a record exemplifies a disruption of radio norms and is rather unusual for light entertainment programmes. The second interruption is more traditional for commercial radio: an advert. At first, there is a sensual female voice that says 'no one is bored with Jean-Bernard' over the last notes of the record. Then a male voice reads an advert for a new vacuum cleaner, with a focus on the fact that it can be tested before purchase. After this advert, the host introduces a new record by the Rolling Stones, and once it is finished, he calls listeners to 'turn up the volume of their transistors until 10 pm'. The recording ends soon after the first notes of the next record, by Sylvie Vartan.

Top 20, 1966, Radio Luxembourg (English Service). The third recording is from Radio Luxembourg's *Top 20*, presented by Barry Alldis, in 1966.[38] This day was actually Alldis' last time presenting it, after being the host for roughly eight years, which might explain why a recording of it was made at the time. The sound quality varies through the recording and the last twenty minutes of it are rather poor, which seems to reflect what was heard on the radio by listeners. In a similar fashion to Brian Matthew's show mentioned above, the *Top 20* was a hit-parade programme, centred on the disc-jockey's personality and on the records played.[39] According to the music magazine *New Musical Express*, the show's sponsor, which first started publishing in 1952, the ranking is based on the best-selling records in Britain. The recording itself opens on a clear sonic icon—a gong—marking the beginning of the programme, which is followed by the enthusiastic voice of the host, Barry Alldis, saying: 'This is your station of stars, Radio Luxembourg!',

[38] Top 20, presented by Barry Alldis, 26th June 1966, Radio Luxembourg 208, 00:56:12, personal archives of Stuart Busby.

[39] In this case, the recording includes, among others, 'Paperback Writer' by The Beatles, 'Strangers in the Night' by Frank Sinatra and 'Monday, Monday' by The Mamas & The Papas.

and one of the stations' jingles, an instrumental of joyful music, mostly played by brass and drums. After this sonic identification of the station, the host introduces the show through its sponsors: 'A leader of Paris, Vienna, London, the people who make Sunsilk Shampoo, Sunsilk Hairspray, Seawitch, Harmony, Melody, Twinkle and Pin-up, [inaudible] the makers of a new ['booping' sound] fried bars, picnic, crunchy, turkish delight and chocolate cream, invite you to Top 20 [lingers with excitement on the word twenty]! Yes, in-dee-dee. And it's a big 'hi' from your D.J. B.A.,[40] Barry Alldis, welcome to the show. Nice to have your company for another sixty minutes of hit-parade programme based on the best-selling records in Britain as published by the *New Musical Express*. How are you tonight? Alright? Good. That's the idea. [...] Oh, by the way [his tone goes deeper, his pace slower and overall, more serious] this is the last time I will be introducing the show. Well, personally, I was a resident Luxembourg disc-jockey,[41] I am off to London to freelance after nine and a half years here, in the Grand Duchy; that is a long time. Next week, Don Wardell, best boy, will be in the chair, we wish the best of luck. Let's not get all sad and sentimental. Let's make the show a swinger and here comes that chart topper! [his voice is highly enthusiastic again in the last sentence]'. The show then follows a straightforward pattern: a record is played by the host, and after a few seconds he speaks, quickly, over the music to name the song and the artist, and stops, just before the lyrics begin. As soon as the record is over, Barry Alldis makes a brief[42] comment about the record that was played and the one to come, often mentioning their ranking on the Top 20,[43] then plays another record, waiting a few seconds before introducing it, etc. and the pattern repeats. Some of the comments include addresses to the listeners, nicknamed the 'Top 20 tuners'. Halfway through the programme, the host repeats the opening advert for shampoo and sweets, and once more

[40] 'D.J. B.A.' was a known on-air catchphrase of Barry Alldis. Janet Alldis, *Under the Bedclothes* (Mersch: Ferall, 1993), p. 128.

[41] The English service of Radio Luxembourg had hosts living in Luxembourg and working in the studios of the Villa Louvigny, while others were recording in the London studios, on 38 Hertford Street, therefore explaining why Barry Alldis made this specification of being a 'resident Luxembourg disc-jockey'.

[42] On average, the comments between the 20 records last less than eight seconds.

[43] Contrary to most contemporary hit-parades, Alldis started with the best-selling record and finished with the twentieth.

in his closing remarks after the last record has played. There is a contrast between adverts and comments, in terms of content. The latter are clearly scripted and lengthier, while the former sound more spontaneous. Both however have a similar inflection with Alldis' regular, fast, and enthusiastic voice. This blurs the limits between programme content and adverts, integrating the commercial aspects into the whole programme.

The three recordings analysed here act as an introduction to the sound of commercial radio in the 'High Sixties', and highlight some of its key characteristics, especially the lasting omnipresence of popular records as a central feature. Indeed, through these recordings, a transnational circulation of records is seen, notably in French-speaking programmes, where French *chanson* coexisted alongside the *yéyé* movement and American and British records.[44] Lastly, another striking aspect is the importance of radio personalities, which is particularly present in the three recordings. Although this is not a new aspect specific to the 'High Sixties', it became even more audible in this time period.

3 The 'Late Sixties'

The following group of recordings brings insights into the last period of the Long Sixties studied in this work. The 'Late Sixties' refers to the years 1968 to 1974, and therefore points to the length of the period, which, as explained in the introduction, should be interpreted beyond the strict limits of the decade. In this late period, what was seen as a novelty beforehand became more normal, while elements rooted in the 1950s disappeared from the radio landscape, following the changes of the 'High Sixties'.

[44] For more details about the history of popular music in France in the 1960s and the tensions between *chanson*, *yéyé* and foreign influence, see Jonathyne Briggs, *Sounds French: Globalization, Cultural Communities, & Pop Music, 1958–1980* (Oxford: Oxford University Press, 2015). For a comparative approach between France and the United Kingdom, see Hugh Dauncey & Philippe Le Guern (eds.), *Stereo: Comparative Perspectives on the Sociological Study of Popular Music in France and Britain*, 2nd ed. (Abingdon: Routledge, 2016).

Hubert, 1970, Europe n°1. The first recording is an extract from a record show presented by Hubert Wayaffe[45] on Europe n°1 in 1970,[46] simply called *Hubert*. He was a French radio host, who worked on Europe n°1 on multiple programmes where he tried to bring to the air the American concept of disc-jockey.[47] This particular show was pre-recorded—and later played on the air[48]—in Tahiti, in French Polynesia, as the host mentions a few times: 'I hope that this little show I prepared for you from Tahiti, and which I present from Tahiti, in Polynesia, suits you. In my village at the end of the world, it is actually Joe Dassin's song that [...] I am talking to you from my village at the end of the world, on Europe n°1, my village is called Tahiti [a record starts], it's a small island, you will see on the map where it is [...] I am sure that some of you might even wish to come to Tahiti, it's easy, you tune in to Europe n°1 every Sunday, and, well, you listen to me, my name is Hubert. You can even send me a little note, I will send you a postcard from Tahiti, they're very nice [...] It is exactly like you picture it—Tahiti—but a little bit more beautiful, *voilà*'.[49] The show is made of records played and chosen by the host, interspersed by his comments announcing titles and artists. At times, Hubert would also include the way to reach him, at 'Postal

[45] Hubert Wayaffe (1938–2019) was a French disc-jockey and radio host. He joined Europe n°1 in 1963, where he started on the night programme, before collaborating on other shows, such as *Dans le vent* and *Salut les Copains*. See 'Hubert Wayaffe, animateur emblématique de 'Salut les Copains', est mort', 13th April 2019, Europe1.fr, [https://www.europe1.fr/medias-tele/hubert-wayaffe-animateur-emblematique-de-salut-les-copains-est-mort-3891065] (last accessed 3 March 2023) and Franck Ferrand, *Au coeur de l'histoire*, 14th September 2012, Europe 1, show dedicated to *Salut les Copains*, with Daniel Filipacchi and Hubert Wayaffe as guests, [https://www.europe1.fr/emissions/Au-coeur-de-l-histoire/L-INTEGRALE-Au-coeur-de-l-emission-Salut-les-Copains-98248] (last accessed 3 March 2023).

[46] The exact date is unknown, apart from the fact that it is broadcast on a Sunday afternoon.

[47] D. Olivennes, F. Ferrand & B. Labous, *Europe 1. Dictionnaire amoureux illustré* (Paris: Plon, 2015), p. 211.

[48] Contrary to the norm, Wayaffe did not pretend his show was live, fully admitting it was pre-recorded. It is possible that he knew people would not believe he would broadcast live from an island so far away, or that the exotism of Tahiti would make up the pre-recording. As a result, this downgrades the feeling of liveness of this programme.

[49] *Hubert*, 1970, Europe n°1, 00:31:34, Archives of Europe 1, Paris.

Box 150, Paris 8th'.[50] Throughout the show, there are no moments of silence or pauses between records and comments by the host, who speaks on the first and last seconds of each song, blurring the lines between the various sonic elements of the programme. One key example is at the beginning of the recording, when Hubert speaks while 'Fortunate Son' by Creedence Clearwater Revival plays. Specifically, he says 'hello everyone'; 'This is Hubert speaking to you from Tahiti'; 'on Europe n°1' between the famous guitar riffs of the intro, when only the bass and drums are playing, strongly entangling the host's voice with the records played. The records played during the show are a mix of American and French artists: Sylvie Vartan, Joe Dassin, and Barbara are played as well as Bob Dylan and Creedence Clearwater Revival. The host also pronounces English names and titles with almost no French accent and even includes catch-phrases in English here and there (*i.e.* 'Let's go!'), which sounds unusual in comparison to other French radio recordings of the period. In the available extract, there was no advertisement, however, this does not mean there were none during the programme.

RTL Non-stop, 1972, RTL. The second recording is an extract from *RTL non-stop*, starring Tino Rossi[51] that was broadcast on the eponym station in 1972.[52] This show was inspired by the music-hall format and presented by Philippe Bouvard,[53] one of the most famous French voices of commercial radio and light entertainment. In *RTL non-stop*, Bouvard, as the host, invited artists, mostly singers, to discuss and perform songs in front of a live audience. In this recording, the guest is the French singer Tino Rossi. The recording, part of the *Fonds Gaston L'Herbier*, is only an extract of a longer show. It opens in the middle of Rossi's 'Pluie d'été et soleil d'hiver', which is followed by a discussion between the host and the singer. Bouvard mentions that the studios are 'a little bit [like] the

[50] 8th being a reference to the *8ème arrondissement* de Paris, the district of Europe n°1 studios on Rue François 1er, the same address given by Daniel Filipacchi in *Salut les Copains*.

[51] Constantin 'Tino' Rossi (1907–1983) was a French singer. Born in Ajaccio (Corsica, France), he recorded a particularly large number of songs, his most famous hit likely being 'Petit Papa Noël'. Rossi also had an acting career.

[52] *RTL non-stop* (with Tino Rossi), 1972, RTL, 00:31:08, INAthèque, Paris.

[53] Philippe Bouvard (born 1929) is a French radio host and journalist, who worked for the *Figaro* newspaper in the 1960s, as well as for Radio Luxembourg, where he started his radio career as a reporter. From 1967 to 1974, he hosted *RTL non-stop*.

atmosphere of your homeland', to which Rossi replies that people are actually working in the studios, followed by Bouvard then corrects him, and explains that he was referring to the heat. Laughs from both men and the audience conclude the exchange. The programme then proceeds with a series of songs interpreted by the guest, with the strong audio presence of the audience, which is often heard cheering, applauding, and even singing the chorus at times. After four songs, the host announces that the next song is new, which generates a clear excitement within the audience. The new song performed by Rossi is called 'Parle Plus Bas' and is a cover, in French, of 'Speak Softly, Love' by Andy Williams, itself a rendition of an instrumental theme of Francis Ford Coppola's 1972 film *The Godfather*.[54] The recording includes four more songs, performed by the guest singer, and each one is met with enthusiasm and applause from the audience. This recording is especially noteworthy for the omnipresence of the audience. A key characteristic of the programme is that it tried to recreate a music-hall/cabaret atmosphere through its elements of liveliness, such as people cheering and talking, which provide the listener with an immersive experience. The presence of the live audience also adds a strong emotional layer to the audio source, through their reactions to songs, especially, to the host's announcement of a new song by Rossi, and this new layer is a fascinating modulation to a musical performance on radio.

The Esso Show, 1970, Radio Luxembourg (English Service). The third recording is from the *Esso Show*, a record show, sponsored by the petrol company Esso, and presented by Kenny Everett,[55] on Radio Luxembourg 208 in November 1970.[56] Likely recorded by a listener, the

[54] It might be relevant to mention that Tino Rossi speaks French with a Corsican accent, not far from Italian and Sicilian accents, quite present in the motion picture.

[55] Kenny Everett (1944–1995) was a British radio and television host, famous for his eccentric and irreverent on-air persona. He began his career on the pirate station Radio London, from where he was dismissed after religious comments on the air. Radio London was indeed partly financed through the broadcast of a half-hour religious commentary, The World Tomorrow, sponsored by an American evangelist church. Everett also worked for the English service of Radio Luxembourg and was among the first to work on BBC Radio 1. He only worked there for a few years, and was let go after a joke about a Minister's wife, which was deemed inappropriate. He continued his career on commercial radio and television afterwards.

[56] *Esso Show*, presented by Kenny Everett, Radio Luxembourg 208, 13th November 1970, 00:29:07, personal archives of Mick Capewell.

document includes the last moments of the show preceding Everett's, in which its host, Paul Burnett,[57] signs off quickly, with a deep voice: 'And don't forget that, hmm, 208 sound survey is brought to you by Baby Doll Cosmetics and Levi's stores. This is Paul Burnett hoping you have enjoyed this show. I won't be back next Sunday [...] but you can join Kid Jensen[58] for another excitement in radio sound, from 11 o'clock to 12.15 with, of course, the 208 sound survey show. I will be on my way to New York for a short holiday, and I will be back, hmm, the week after I hope. Anyway, don't forget, next Sunday night, you will hear this song and it means you're gonna have a good listening on the trend-setting, face-making, 208 sound survey!'. Proceeding his sign-off is Radio Luxembourg 208's jingle, instrumental and pompous, with many brass instruments. A second jingle follows this time with a man and a woman singing over a joyful and jazzy theme: 'from North to South, from East to West, the folks who want the best, listen to Luxembourg, the great 208, the [unrecognised word] swinging station of the stars'. As soon as the jingle ends, an advert for Esso, the sponsor, is heard; a nursery rhyme, sung by a very young child, which goes as follows: 'The Esso sign means good motoring', repeated multiple times in a short period. The host of the *Esso Show*—Kenny Everett—picks up on the ad and in a serious almost ceremonial manner introduces his show. Next may be described as a parody of the original advert for Esso. It is a chorus, presumably with Everett's voice recorded multiple times, singing about the petrol brand in an almost sexual manner, as if in a love song: 'Here's a petrol that we all love the best. Passed every test. Beaten the rest. It makes me weak [pause] at the knees', to which the host adds, as if talking to himself: 'Must stop drinking it'. This example highlights a key feature that runs throughout the programme: the irreverence towards radio norms. On a few occasions,

[57] Paul Burnett (born in 1943) is a British disc-jockey, who started his career on the offshore station Radio 270, before joining Radio Luxembourg in 1967, where he presented the *Top 20* programme until 1974, when he joined BBC Radio 1. He is also known to be the one who discovered many recordings that were presumed lost, from 'Lord Haw Haw', when he was broadcasting Nazi propaganda to Britain when Luxembourg was occupied during World War II. Nichols, *Radio Luxembourg: The Station of the Stars*, p. 53.

[58] David 'Kid' Jensen (born in 1950) is a Canadian-born British radio host, who joined Radio Luxembourg in 1968 at the age of 18, hence his nickname. He later worked for ITV and BBC Radio 1.

the host seems to make fun of the sponsor, obviously faking his enthusiasm for the collectible gifts available at Esso stations.[59] For example, Everett recommends saying hello to Esso dealers in a highly sexual way, as he presents the available gifts, supposedly so wonderful that he has a series of three loud moans, imitating an orgasm, on air. He also uses a fake Scottish accent at some point and pronounces a few words in what can only be described as a 'silly' voice. This segment is interesting as it shows that, at least in its content, the commercial aspects of the programme are kept and respected, but also twisted and not taken seriously. Another unusual moment regarding radio norms happens fifteen minutes into the recording, when the host imitates a telephone ringtone. A man is on the phone, and says (while Everett yawns in the background), 'Good evening, Radio Luxembourg listeners, this is Jonathan King,[60] broadcasting, live, from the heart of Bayswater, in London. If you like my record "Cherry, Cherry", please go and get a copy: 'cause I need the bread, you see'. In response to King, Everett states that his record sounds better at a high speed; which he proceeds to prove by actually playing it at a high speed, before slowing the record down to silence. Jonathan King, still on the phone, thanks Everett for playing his music, and the two exchange a few compliments, then hang up. This particular interaction is a good example of playing with radio norms, as it is unusual for disc-jockeys to play records in such a way that makes them inaudible. The programme otherwise follows the traditional format of a record show on this station, with records[61] and adverts interspersed with the host's comments. One particularity of the recording is the importance of sound effects. Electronic sounds to punctuate sentences, echoes in the voices, and honking noises are all part of the technical arsenal used by Everett to modulate his programme, to an extent not met in the other shows studied in this section.

One striking element that appears from the analysis of the three recordings presented here is the advent of the central figure of the disc-jockey in commercial radio. Indeed, both Hubert and Everett clearly put their

[59] The gifts were called Esso Book of Squelchers; a series of collectible miniatures booklets containing information about football.

[60] Jonathan King (born in 1944) was a British singer-songwriter and record producer, who was also involved in television and radio activities.

[61] Among others: 'Dismal Day' by Bread, 'Think About Your Children' by Mary Hopkins and 'Fire And Rain' by James Taylor.

on-air personalities at the centre of their respective shows, and it is their perceived personal traits which 'made' the programmes. This evolution can easily be seen elsewhere in the period and was already quite prominent in the British context. Nevertheless, RTL's programme offers a counterexample as the cabaret-inspired format of Bouvard's show,[62] the presence of a live audience, echoes older formats. This is highlighted by the slow evolution of radio programming, as well as its variety, as 'older' formats cohabited with more 'modern' examples, showing the fluidity of the sound of commercial radio in the Sixties.

These textual descriptions of the chosen recordings are helpful to get a sense of what commercial radio sounded like in the Sixties. Based on these nine audio sources, some analysis can be formulated, but the importance of music contemporaneous to the time period should be noted. Popular records are central to most of the programmes studied, particularly when studying the English service of Radio Luxembourg. There are some differences that are to be expected between programmes in French and English, and their changes are reflected in the evolution of the Long Sixties and the changes in popular music. The records are often linked to rankings and so-called hit parades, based on partnerships with magazines, such as *The New Musical Express* for Radio Luxembourg's *Top 20*, and on internal calculations for *Salut les Copains*. What clearly changed over the course of the decade is the place of these records within the programmes; in *Pour ceux qui aiment le jazz* and in *Your Record Show*, there are pauses between records and other elements of the radio programme. This is something that disappears in later shows, wherein the elements are blended together; the recurrent case of the host speaking over the first and last notes of a record exemplifies one such evolution. It also echoes the growing reduction of silences and pauses in radio programmes throughout the Long Sixties. In Kenny Everett's *Esso Show*, the host speaks particularly fast and modulates his voice to make a variety of human sounds (*i.e.* moaning) and electronically generated sounds (*i.e.* honking) in addition to his own speech, something that was not present earlier in the period. It is important to mention that radio programmes are often linked to one another. These connections are especially striking when assessing sources recorded by amateurs, in which the recording does not necessarily fit the timeslot of the programme on air. For example,

[62] Moreover, Tino Rossi's career began long before World War 2, making him, the guest, and, by extension, the show, rather 'old-fashioned'.

one host may be heard announcing the following show or thanking the previous host. These audio-connecting elements are important as they link together the various programmes of a station, which are often seen as independent entities due to the way radio is produced, as well as archival policies and timetables of shows in magazines. These connections also occur over longer durations—one such example being Daniel Filipacchi playing a record that he had an issue with the previous day—connecting the shows over different days. These elements appear important in order to understand that radio programmes did not exist as individual moments or entities, despite being presented as such in the sources. Instead, they are better understood as being linked to one another throughout the flow of radio.[63]

References

Alldis, Janet. *Under the Bedclothes*. Mersch: Ferall, 1993.

Arnheim, Rudolph. *Radio*. Translated by Margaret Ludwig & Herbert Read. London: Faber & Faber, 1936.

Birdsall, Carolyn. *Nazi Soundscapes: Sound, Technology and Urban Spaces in Germany, 1933–1945*. Amsterdam: Amsterdam University Press, 2012.

Briggs, Jonathyne. *Sounds French: Globalization, Cultural Communities, & Pop Music, 1958-1980*. Oxford: Oxford University Press, 2015.

Chignell, Hugh. *Key Concepts in Radio Studies*. London: Sage, 2009.

Chion, Michel. *Le Son. Ouïr, écouter, observer*, 3rd ed. Malakoff: Armand Colin, 2018.

Chion, Michel. "100 concepts pour penser et décrire le cinéma sonore". Online, October 2012. [http://michelchion.com/]

Cohen, Debra Ray, Michael Coyle and Jane Lewty, eds. *Broadcasting Modernism*. Gainesville: University Press of Florida, 2009.

Connor, Steven. "Edison's Teeth: Touching Hearing". In *Hearing Cultures. Essays on Sound, Listening and Modernity*, edited by Veit Erlmann, 153–172. Oxford: Berg, 2004.

[63] The concept of flow, applied to radio, can be defined, in a few words, as 'a single irresponsible flow of images and feelings' (Raymond Williams, *Television: Technology and Cultural Form* [New York: Schocken Books, 1975], p. 91), and was later nuanced by Gripsrud (Jostein Gripsrud, *Understanding Media Culture* [London: Arnold, 2002]) and Ellis (John Ellis, *Visible Fictions. Cinema, Television, Video* [London: Routledge, 1982]) in a less negative, more neutral way. It is an important concept to apprehend the succession of programmes in media, and their existence as a flow rather than the sum of individual shows.

Crook, Tim. *The Sound Handbook*. Abingdon: Routledge, 2012.
Dauncey, Hugh and Philippe Le Guern, eds. *Stereo: Comparative Perspectives on the Sociological Study of Popular Music in France and Britain*, 2nd ed. Abingdon: Routledge, 2016.
Ellis, John. *Visible Fictions. Cinema, Television, Video*. London: Routledge, 1982.
Erlmann, Veit, ed. *Hearing Cultures. Essays on Sound, Listening and Modernity*. Oxford: Berg, 2004.
Ferrand, Franck. "Au coeur de l'émission Salut les Copains !". *Au Coeur de l'Histoire. Europe 1*, September 14th, 2012.
Goodale, Greg. *Sonic Persuasion: Reading Sound in the Recorded Age*. Urbana: University of Illinois Press, 2011.
Gripsrud, Jostein. *Understanding Media Culture*. London: Arnold, 2002.
Hilgert, Christoph. *Die unerhörte Generation: Jugend im westdeutschen und britischen Hörfunk, 1945–1963*. Göttingen: Wallstein, 2015.
Hubert Wayaffe, animateur emblématique de 'Salut les Copains', est mort. *Europe 1*, April 13th 2019. [https://www.europe1.fr/medias-tele/hubert-wayaffe-animateur-emblematique-de-salut-les-copains-est-mort-3891065]
Jehle, Anna. *Welle der Konsumgesellschaft: Radio Luxembourg in Frankreich, 1945–1975*. Göttingen: Wallstein Verlag, 2018.
Le Breton, David. *Éclats de voix. Une Anthropologie des voix*. Paris: Métailié, 2011.
Maréchal, Denis. *RTL. Histoire d'une radio populaire: De Radio Luxembourg à RTL.fr*. Paris: Nouveau Monde Éditions, 2010.
Nichols, Richard. *Radio Luxembourg: The Station of the Stars*. London: W.H. Allen, 1983.
Obituary: Brian Matthew. *BBC News*, April 8th, 2017. [https://www.bbc.com/news/entertainment-arts-39513249]
Olivennes, Denis, Franck Ferrand and Bruno Labous. *Europe 1: Le Dictionnaire amoureux illustré*. Paris: Plon, 2015.
Prot, Robert. *Précis d'Histoire de la Radio et de la Télévision*. Paris: L'Harmattan, 2007.
Prot, Robert. "Salut les Copains". In *Dictionnaire de la radio*. Grenoble: Presses Universitaires de Grenoble, 1999.
Smati, Nozha and Anne-Caroline Fiévet, "A la radio, la voix donne à écouter et à voir", 04/12/2017, INAglobal.fr [http://www.inaglobal.fr/radio/article/la-radio-la-voix-donne-ecouter-et-voir-10010]
Suisman, David. "Introduction: Thinking Historically About Sound and Sense". In *Sound in the Age of Mechanical Reproduction*, edited by David Suisman and Susan Strasser, 1–12. Philadelphia: University of Pennsylvania Press, 2010.

Tinker, Chris. "Shaping 1960s Youth in Britain and France: Fabulous and Salut les Copains". *International Journal of Cultural Studies* 14:6 (2011): 641–657.

Williams, Raymond. *Television: Technology and Cultural Form*. New York: Schocken Books, 1975.

CHAPTER 6

Key Features of Commercial Radio in the Long Sixties

1 Introduction

This chapter analyses a wider corpus of audio sources. By doing so, common characteristics—already present in the previous chapter—appear and are further developed. First, commercialism is central to the studied stations and makes up the very fabric of the radio programmes through sponsorships, adverts, and links to other media, such as magazines. Second, irreverence is a key characteristic of commercial radio of this period, something that is particularly striking in Kenny Everett's *Esso Show* on Radio Luxembourg. However, it should be added that irreverence in several instances is also linked to a form of playfulness, which is also developed here. Third, interactivity is a strong component of commercial radio, whether it is the presence of an audience at a musical performance, phone-ins by listeners to request songs, or invitations by radio hosts to interact with them and the stations by mail and phone. Fourth, liveness plays a crucial role in commercial radio that needs to be further developed. Last, commercial radio stations have developed an arsenal of sonic icons that constitute an essential point of study of their sound. The last section is devoted to analyse the ways through which sound gives an indication of the materiality of radio.

R. Legay, *The Cultural Histories of Radio Luxembourg and Europe n°1*,
https://doi.org/10.1007/978-3-031-46250-4_6

2 Commercialism

The radio landscape in the Long Sixties is strongly divided between public and private broadcasters, also known as commercial radio stations. It is therefore necessary to start this chapter by looking into this feature. Commercialism is heard on air through multiple elements, in particular adverts and sponsored programmes, results of partnerships between the stations and various businesses. The importance of advertisement in the sound of the stations is of central importance. This is obviously a strong characteristic of commercialism in radio.[1] Adverts can be woven into the programme, as is the case with Kenny Everett's show, in which the host integrates the sponsor, Esso, into his own speech, thereby playing with the norms of commercial radio. In other cases, adverts act as moments of separation; they are read by the host, or by another voice, and create pauses within the programme. An extreme example is Horace Batchelor's particularly long and slow advert for the Infra-Draw method, which sounds different from the rest of the programme. It may even be argued that the ad creates an anachronic feel for listeners.[2] French-speaking programmes operate in a similar way, with adverts being read by another voice, often female, or by the main host. Some adverts are also integrated into the programme in a more subtle way, such as ads by sponsors, which have a prominent place in the show. Adverts are omnipresent in the sound of commercial radio, yet their content, often dissonant from the programmes', is something radio hosts tend to overlook. Presumably, it was a strategy to smoothen the continuity of their shows. Occasionally though, hosts would question the commercial aspect. This is exemplified by the evening programmes of RTL on 26th September 1968. One of the co-hosts, Gérard Klein, explains that the record he plays—an instrumental Gypsy jazz song—was written as a homage to French actress and singer Brigitte Bardot, a comment that makes his co-host, Nicole, wonder

[1] The links between advertisement and radio have been particularly well studied in the United States, where 'as a new and influential component of a consumer society, American broadcasting grew throughout the 1920s into a highly profitable industry, largely on the sale of airtime for advertisements'. See Michael Stamm, *Sound Business: Newspapers, Radio, and the Politics of New Media*, 2nd ed. (Philadelphia: University of Pennsylvania Press, 2016), p. 40.

[2] Batchelor's speech echoes those of radio presenters of the 40 s and 50 s due to its slow pace, clear enunciation, and choice of words. Coupled to the man's age, it creates this anachronic sensation.

if it is not strange to move on from this to an advert for cornmeal cakes. Reassured by Klein, she enthusiastically carries on with her advert praising the qualities of Alsa cornmeal cakes, while the gypsy jazz keeps playing, faintly, in the background. There are instances where adverts are mocked, as was the case in Kenny Everett's show, slightly turning advertisement—a key feature of these stations—into derision. There is indeed a tension between radio programmes and adverts, which can sound out of place, or even, as pointed out by Nicole, absurd, although the stations seem to work hard to make this tension as discreet as possible.

Commercialism of the studied stations also extends to a more individual level, in particular, the case of Daniel Filipacchi, who often references the *Salut les Copains* magazine in the eponymous show and the *Mademoiselle Âge Tendre* magazine. His references blur the lines between self-promotion and the station's connection to the press, as he owned numerous magazines. In the aforementioned example, he invites participants of a beauty pageant organised by *Mademoiselle Âge Tendre* on his show *Salut les Copains*.[3] Amusingly, in the same programme, his tongue twists and he mentions the listeners of 'Mademoiselle Âge... Salut les Copains'. Although the relationship between radio and magazines is studied at length in the third part of this book, it is worth briefly noting here due to the clear demonstration of their connectedness made by Filipacchi, whose voice and position at the heart of a small transmedia empire, used commercial ventures for personal advancement.

3 Irreverence and Playfulness

One component that seems to apply to all commercial radio stations is a taste for irreverence and playfulness, which has its roots in the station's longstanding pride of their supposed freedom. Mostly constructed in opposition to public broadcasters, whether the BBC or the (O)RTF, this image is based on the dichotomy between a public service broadcaster and a commercial enterprise, the latter being seen as having more liberty to be pleasant. Playfulness on radio has a long history,[4] and it remains present in

[3] Salut les Copains, Europe n°1, 1966, 01:08:00, Archives Europe 1, Paris.

[4] See, for instance, Maria Rikitianskaia, 'A Transnational Approach to Radio Amateurism in the 1910s', in G. Föllmer & A. Badenoch (eds.), *Transnationalizing Radio Research: New Approaches to an Old Medium* (Bielefeld: Transcript), pp. 133–140. Moreover,

the Long Sixties, but it also coexists with a strong tendency towards irreverence,[5] which was already illustrated in the *Esso Show* by Kenny Everett, described in the previous chapter. These two ideas include moments when radio hosts are bending radio norms for amusement purposes. At times, such ventures manifest in more extreme manners, *i.e.* jokes, actions, and comments that might easily be understood as disrespectful. Irreverence, especially in regards to adverts and sponsors, regularly takes the shape of sexual connotations. Everett was mentioned, although Gérad Klein does something similar when he mocks the stockings in an advert, presented by his co-host Nicole. Specifically, he makes sexual sounds when Nicole refers to stockings. Perhaps it is an attempt to be playful and distract her, but it is nevertheless questionable behaviour considering the importance of sponsors for commercial radios.

Direct playfulness can be heard in a recording of Pierre Bellemare, from an unidentified extract. Taken from a source at the INAthèque, the extract offers a mix of programmes from 1966 on RTL, including a game during which Bellemare calls random phone numbers. To the surprise of everyone in the studio, the host ends up on the phone with the answering machine of a sex line. A female voice answers and provides him with the name Frederica, her mensuration, rates, and address.[6] Multiple laughs are heard in the studio, while the host jokes about writing down the address, likely in order to regain control of the programme's narrative. Unlike playfulness, irreverence often goes a step further. One example can be found in a *Balzac 10–10* show, wherein the host is forced to interrupt his speech, as he is taken into a fit of uncontrollable laughing. Feeling

Spinelli, in his analysis of Wells' *World of Wars* showed that 'far from ending the possibilities for the radio prank, *War of the Worlds* ushered in radio's trickster aesthetic in both high and low cultural spheres', which echoes the irreverence and playfulness studied in this section. See Martin Spinelli, 'Masters of Sacred Ceremonies', in D. R. Cohen, M. Coyle & J. Lewty (eds.), *Broadcasting Modernism* (Gainesville: University Press of Florida, 2009), pp. 68–88.

[5] This irreverence might find some roots in the culture of pirate and offshore radio stations. For instance, Ronan O'Rahilly (1940-2020), who launched Radio Caroline, through a purposefully complex and confusing scheme, shaped his own image as a rebel. He built it, among other things, on his grandfather, Michael Joseph O'Rahilly, a key figure of the 1916 Easter Rising in Dublin, and on the fact that he was—allegedly—expelled on seven occasions from various schools. See Ray Clark, *Radio Caroline: The True Story of the Boat that Rocked* (New-York: The History Press, 2014), p. 33.

[6] Compilation d'extraits de RTL, RTL, 1966, 00:31:08, INAthèque, Paris.

the need to explain this interruption to listeners, he explains that a team member is laying on his back on a table and sleeping, or pretending to sleep, in front of him,[7] disrupting the programme through unprofessional behaviour. Another example of mischief from radio staff whose behaviour impacts the host's speech—although the action is not directly present on air—occurs in the studios of Europe n°1. As a radio host tries to read an advert, he is forced to stop due to laughter and explains he cannot focus because of 'clowns' behind a glass window in the studios.[8]

Playfulness can also be a defining characteristic of a programme. In 1961, a show had the singer Henri Salvador appear as a guest[9] and used his loud and iconic laugh to create a jingle, which is then used to tease him. Later in the programme, the hosts play two songs and then proceed to make fun of, due to their rather poor quality. Finally, a third song is a parody, sung by Jean Yanne, mocking young male rock singers, who are often heard on the airwaves of the station. One can also find a show, presented by Francis Blanche, in which he only does phone pranks to random people.[10] Another example, on Europe n°1, is a show called *Equipe n°1*,[11] in which listeners are asked to send bad lyrics to the station for the hosts to interpret on air. It also includes a series of sketches with irreverent aspects, such as racist accents, jokes about French singer Dalida and her talent, a parody of a cooking show for cannibals and a sketch about the Volapuk language (invented in the nineteenth century to be used internationally), the host joking that is the language that should be used by Europe n°1.

Irreverence and playfulness are used to different degrees in several ways, but they appear as strong characteristics of the commercial stations. In a common joint in which RTL and Europe n°1 parody themselves, which is studied later in the next chapter, this irreverence is overt, and acts as a bonding characteristic between the two *radios périphériques*.

[7] Balzac 10-10, Radio Luxembourg, 1963, 00:15:41, INAthèque, Paris.

[8] Fou-rire lors d'une publicité, Europe n°1, 1969, 00:01:36, Archives Europe 1, Paris.

[9] Ah ah ah, Europe n°1, 1961, 00:10:12, Archives Europe 1, Paris.

[10] Canular téléphonique, Europe n°1, 1961, 00:01:24, Archives Europe 1, Paris.

[11] Equipe n°1, Europe n°1, 20th May 1962, 00:55:51, Archives Europe 1, Paris.

4 Interactivity

Whether audiences are listeners or game participants, many programmes of commercial radio stations pay particular attention to the relations with their audience, which helps to develop a strong sense of interactivity. This interactivity occurs via a variety of exchanges, which can take many forms, including phone-ins, audience participation in radio games[12] and competitions, hosts directly addressing their listeners, the letters sent to the stations and sometimes read on air, etc. In other words, the development of interactivity is significant as it moves the audience into a more active position, while having them sonically present, regardless of whether they speak, are read about, or simply mentioned.

Phoning-in is a technique that allows listeners' calls to be incorporated into a radio programme.[13] Phone-ins are common in a large number of programmes on commercial radio stations, especially in game shows of French-speaking stations, for the very reason that it readily allows listeners to ask questions to radio hosts and their celebrity guests. An early occurrence found in the recordings dates back to 1959[14] in a programme on the French service Radio Luxembourg, in which three older women share their personal wishes. One of the women's is picked by listeners to be fulfilled by the station. The winning wish is decided upon by three groups of listeners, formed in advance and gathered in the same place for the show, and called by Radio Luxembourg. One woman's wish was to visit the studios of the station. Although her wish does not win, the programme receives a phone call while on air. A member of staff transfers the call, and it turns out to be a message from one of

[12] Radio game shows, seen as an ideal type of programmes for interactivity with listeners and live audiences, have known their golden days in France after World War 2. However, as shown by Marie-Paule Schmitt, they faced a strong decline from the 1960s onwards, slowly disappearing from airwaves. It is, nevertheless, important to nuance this disappearance, as they actually moved to television, which sometimes even kept the same teams and barely adapted the format to the new medium. See Marie-Paule Schmitt, 'Les Jeux sur les ondes de la Libération aux années soixante-dix: grandeur et décadence d'un genre radiophonique', *Sociétés & Représentations* (24:2, 2007), pp. 353–369 & Marie-Paule Schmitt, 'Les Jeux radiophoniques, en France, dans les années soixante', *Sociétés & Représentations* (37:1, 2014), pp. 219–228.

[13] According to Chignell, the first British phone-in occurred in 1968, whereas the technique seems to have been used on Radio Luxembourg and Europe n°1 earlier. Hugh Chignell, *Key Concepts in Radio Studies* (London: Sage, 2009), p. 37.

[14] Le rêve de votre vie, Radio Luxembourg, 1959, 00:31:00, INAthèque, Paris.

the directors, Matthias Felten, who explains that he was listening to the programme, and wished to invite the participant to visit the studios in both Paris and Luxembourg, and the Grand Duchy. This genuine aspect of Felten's message is possibly staged, however, the programme interestingly presents listeners from both ends of the spectrum, with the unnamed regular listeners and the station's director. Another aspect of interactivity that affects the sound of commercial radio, in this programme, is the dichotomy between the voices of audience participants and radio professionals, as they sound deeply different. In the aforementioned game, three older women are interviewed (their older age is indicated by the sound of their voices) and two of them have heavy accents, from Burgundy and Normandy. They all share some memories, such as their first encounter with radio,[15] their job in a *grand magasin,* and their favourite pet goat, which are all elements that infer a popular background, contrasting with the professional elocution of the host.

Phone calls were at the core of a game on Europe n°1—*Un millionnaire au bout du fil* (A millionaire on the line)—during which the participant had fifteen minutes to spend a large amount of money by ordering things on the phone.[16] The difficulty is that the show is in the evening when most stores are closed, and their owners are a bit sceptical about massive sales on the phone. Most of the recording entails the participant searching through the phone book, trying to make these phone calls, writing down names and numbers, calling the operator and asking to be put through to specific numbers, and waiting to be called back by the operator. Actual conversations on the phone are quite rare. One short recording from *Salut les Copains* in 1966 focuses solely on a phone conversation between French singer Sylvie Vartan and a young female listener, Dominique. This particular conversation was likely preserved because of the apparent youth of the listener, whose voice and word choice indicate she is between ten and twelve years old. Such a recording

[15] The woman explains that, when she was in high school, in 1911, a man was going from school to school to present Marconi's experiments, something that seems to have a strong impact on the host.

[16] Un millionnaire au bout du fil, Europe n°1, 24th September 1961, 00:24:46, Archives Europe 1, Paris.

offers an unusual voice on radio and helps to create a friendly, warm, interaction between the participant and the singer.[17]

On two separate occasions, the interactivity between Europe n°1 and its listeners takes an interesting form, as it occurs in cars. The first example is during the programme *Dans le Vent* in 1963,[18] in which the host, Michel Cogoni, at roughly one hour into the programme, goes into a Europe n°1 car with his guest, French singer and actress Dalida. As the two head to a record shop, Cogoni commentates their journey on the *Champs-Elysées*, near the Café Georges V, by the *Place de l'Etoile*, etc., anchoring a Parisian vibe to their adventure, while background sounds illustrate a car journey into a big city (*i.e.* sounds of the car engine, honking, doors opening and closing).[19] Another factor to note here is the sound changing between the studios where the show started, to the car, indicating that there was probably a change for a portable one.[20] At one point, the host exclaims: 'Oh, it's wonderful! There is a taxi driver who is very, very nice, who is there, in front of us at the red light, and who's waving at us [...] and letting us know he is currently listening to Europe n°1 and that he hears us at the same time'—to which a neighbouring car, likely the taxi driver's, honks in the near distance, as if he wanted an audio

[17] If phones played an important role for commercial radio, notably in France, they were not so common in the country. In 1965 and in 1970, the equipment rate in landlines in France was slightly behind Italy and West Germany and far lower than in the United Kingdom and the United States. Delays to obtain a line were also particularly high in France. See Pascal Griset, 'Le développement du téléphone en France depuis les années 50. Politique de recherche et recherche d'une politique', *Vingtième siècle, revue d'histoire* (24, 1989), pp. 41–54. See also Marie Carpenter, *La Bataille des télécoms. Vers une France numérique* (Paris: Economica, 2011), p. 41.

[18] Dans le vent, Europe n°1, 17th December 1963, 01:48:46, Archives Europe 1, Paris.

[19] The importance of sounds in relation to urban soundscapes should not be neglected. Sounds can indeed be used in fiction films to clearly identify cities. See Jasper Aalbers, 'Echoes of the City: Staging the Urban Soundscape in Fiction Film', PhD diss., Maastricht University, 2013. Hans-Ulrich Wagner, in a case study analysing the sound of Hamburg, defends the idea that media—notably radio—play a decisive role in the shaping of space-related identities, something which is echoed in the 'Parisian vibe' of the studied show. Hans-Ulrich Wagner, 'Sounds like Hamburg... Hamburg-Klänge und mediatisierte Raumkonstruktionen in der Frühzeit des Radios', in J. Müske, G. Föllmer, T. Hengartner & W. Leimgruber (eds.), *Radio und Identitätspolitiken: Kulturwissenschaftliche Perspektiven* (Bielefeld: Transcript, 2019), pp. 55–72. See also Karin Bijsterveld (ed.), *Soundscapes of the Urban Past* (Bielefeld: Transcript, 2013).

[20] This technical change indicates the importance of technology within the wider *dispositif* of commercial radio stations.

confirmation that the taxi driver had heard them talking about him. This short interaction creates a complex interactivity, as the listener in his taxi is not only mentioned by the host but is also heard reacting to this mention, while it is suggested he is listening live to the programme, including the part in which the show mentions him. Later in the programme, Dalida is in the record shop, where she is tasked with convincing customers to buy her songs. Out of the four who are interviewed, two explained that they had rushed to the shop as they were driving because they heard Dalida was in the neighbourhood. Another had been listening at home and came quickly to the record shop, while the last person saw her in the street and then walked into the store. Having this knowledge reduces the seemingly genuine randomness of the interactions between the celebrity and customers, but at the same time, it shows interactiveness between the station and its listeners, who interrupted their activities to be a part of the physical experience of the radio programme. It points to the power of radio to conjure listeners to depart from the listening experience to join a physical interaction.

In another occasion, *Dans le vent* offers great insight into the interactivity with their listeners. This time it is in 1964 and the guests are French singers Hugues Aufray and Enrico Macias,[21] who are sent out into the streets of Paris on a tandem bike, while the host, Michel Cogoni, follows in one of the station's cars. This setup leads to a particularly chaotic ten-minute-long scene. The bike has neither lights nor brakes, and bad traffic leads the host's car to become blocked. The host then struggles to describe what is happening, until he catches up with the cycling duo, who are said to be helped by 'friends of the station'. The friends push the bike, with 'their transistor over the shoulder', which suggests a group of people using the mobility of the transistor to directly interact with the programme[22] and its guests. As was the case for the show with Dalida, the direct interactions were possible thanks to the geographical mentions made by the host, who describes their whereabouts, such as street names, allowing listeners to join the events described. Towards the end of the recording the group heads back, by car, to the studios. An unknown male voice is heard, in the distance, yelling towards the group

[21] Dans le vent, Europe n°1, 20th May 1964, 00:56:38, Archives Europe 1, Paris.

[22] Richard Legay, 'RTL & Europe n°1 as central actors. The Importance of Mobility for Commercial Radio Stations during the Parisian Events of Mai 68', *Rundfunk & Geschichte* (3:4, 2018), pp. 41–51.

that they were 'not in the right direction for Pigalle!'.[23] This prompts the host, Cogoni, to explain what had just happened to potentially confused listeners: the voice was from a taxi driver nearby, who was listening to Europe n°1 in his car. Replying this time to the driver, the host yells back that they 'are just going for a drive'. This is interesting because, as Cogoni explains, the driver was listening to the station, when he interrupted the programme to give directions. Therefore, there was no need for the host to address him directly; a comment on air would have had the same result. Cogoni, nevertheless, circumvented his radio programme to talk to the driver, reinforcing the power of live broadcasting for interruptions. These two examples are fascinating as they reveal the complexity of interactions between listeners and radio stations outside of the studios and point at the strong anchoring into Parisian life.

These above examples point to interactivity with listeners happening outside of the station's control. The station can also use this spontaneous interactivity purposefully. RTL hosted a show called *RTL dépannage*[24] in which listeners could ask the station to help them fix a problem. For instance, in June 1966, a priest in Picardy, who urgently needed 300 chairs for his church.[25] The host reads the priest's letter on air, which had been sent to the station. He then waits for listeners' calls to come in, offering help. On Christmas Day in 1973, an unusual show aired on Europe n°1. The show was presented by Pierre Bellemare, who explains, on air, that the station received a gigantic *bûche de Noël*,[26] that weighed around 750 kilograms, due to an order cancelled at the last minute; the pastry chefs, having no idea what to do with it, decided to call Europe n°1 and give them the *bûche*, to ensure it did not go to waste. The host explains it was decided to share the *bûche* with listeners and then proceeds to give the address of the studios in Paris, rue François 1er, so people can join the team and share the traditional Christmas treat. It seems the programme also broadcasts exchanges with listeners; however, they are not included in the recording, which stops beforehand. It is another useful source for understanding interactivity; the pastry chefs

[23] 'Pigalle' refers to the *Place de Pigalle* and its surroundings, in Paris.

[24] *Dépannage* could be understood as troubleshooting.

[25] RTL dépannage, RTL, 12th June 1966, 00:20:52, Fonds Gaston L'Herbier, INAthèque, Paris.

[26] A traditional French dessert eaten at Christmas, shaped like a log.

who called the station were likely listeners familiar with Europe n°1, and, by extension, its network of listeners.[27] In both occasions, commercial stations demonstrated reliability and reciprocity between themselves and their listeners to help each other. At the same time, the stations were able to place themselves in a favourable role, as both makers and enablers of good deeds. Europe n°1 even appears to go a step forward on this occasion, as the suggested tasting of the *bûche* happens in the station's studios.

5 Liveness

Liveness is the sense of radio being live, whether it actually is or not,[28] and it is a concept encompassing all elements that provide radio broadcasts with the sense of being live. As such, liveness has been crucial since radio's early days.[29] At the beginning, 'all broadcasting was, perforce, live at first', and it was only later in the twentieth century that recorders allowed to capture language and liveness.[30] It also plays a crucial role in terms of 'keeping people listening',[31] which, in turn, influences the revenue of commercial stations. It can therefore be argued that liveness is more important for understanding private broadcasters as opposed to their public counterparts. To develop this sense of liveness, radio can count on the interactivity, more precisely phone-ins, which were previously developed on liveliness[32] of radio speech and on unusual sonic elements, disrupting the continuity of radio, and supporting the idea

[27] The programme *Vous êtes formidable* is, in this regard, crucial as it shaped the image of Europe n°1 as a 'station with a big heart', by showing how its listeners could be relied on. See Luc Bernard, *Europe 1. La Grande Histoire dans une grande radio* (Paris: Centurion, 1990), p. 98.

[28] Frédéric Antoine, *Analyser la radio. Méthodes et mises en pratiques* (Louvain-la-Neuve: De Boeck, 2016), p. 237.

[29] Alexander Russo reflects on the tensions between the aesthetics of liveness on radio and the presence of recorded programmes, already in the beginning of the twentieth century. See Alexander Russo, *Points on the Dial. Golden Age Radio beyond the Networks* (Durham: Duke University Press, 2010), p. 84.

[30] Paddy Scannell, *Media and Communication* (London: Sage, 2007), p. 184.

[31] Chignell, *Key Concepts in Radio Studies*, p. 90.

[32] Liveliness in broadcast is understood as radio speech giving an impression of spontaneity, unscriptedness. See Andrew Tolson, *Media Talk: Spoken Discourse on TV and Radio* (Edinburgh: Edinburgh University Press, 2006), p. 9.

that the programme is being broadcast live. These elements echo Roland Barthes' concept of *effet de réel*, used mostly in literature, but which can also be used when studying radio sound archives.[33] It is a major aspect of radio as it provides the sense of a shared experience, between the stations and their listeners, and among the listeners.

In a recording of Barry Alldis' *Top 20* mentioned in the previous chapter,[34] it was noted that the disc-jockeys addressed listeners on a few occasions as the 'Top 20 tuners', suggesting they belong to the same group, and also asking them direct questions such as 'How are you tonight?'. Direct questions created a sense of familiarity and community and reinforced this idea of co-presence, which is understood here as the shared experience of radio listening.[35] The inclusion of listeners in the radio speech can take many forms. For example, in one show on Europe n°1 in 1960, Francis Blanche said hello to listeners first, then to another radio host who introduced him.[36] There exists one substantial difference between radio programmes in French and in English, and it is based on the language itself. In English, the pronoun 'you' can address both individuals and groups, whereas French requires the use of 'tu' for individuals and 'vous' for plural and/or formal addresses. The latter, or the plural form, is most commonly used to address listeners as a whole. One occasion in the corpus offers a rather different take on this component of radio speech. In an amateur recording of the evening programmes of RTL in 1968, the host of one of the programmes, Gérard Klein, addresses the listeners in an unusual way: 'Tomorrow, RTL, *Monsieur* Yves Montand, yeah, *Monsieur* Yves Montand, will sing for you'. Referring to listeners as RTL contrasts the usual way they are addressed throughout the corpus, in which the hosts act as representatives of the station. It is important to mention that this is the only time listeners were referred to in this way

[33] Richard Legay, 'The Role of Commercial Radio Stations in the Media Vacuum of *Mai 68* in Paris', *VIEW Journal of European Television History and Culture* (6-12, 2018), pp. 1–11. See also Hervé Glévarec, 'Le reportage radiophonique des 'événements': Mai 68 a-t-il inventé 'l'effet de réel' en radio ?', in Christian Delporte, Denis Maréchal, Caroline Moine and Isabelle Veyrat-Masson, eds., *Images et sons de Mai 68 (1968-2008)* (Paris: Nouveau Monde éditions, 2011), pp. 43–53.

[34] Top 20, presented by Barry Alldis, 26th June 1966, Radio Luxembourg 208, 00:56:12, personal archives of Stuart Busby.

[35] Chignell, *Key Concepts in Radio Studies*, p. 74.

[36] Le Sourire de Francis Blanche, Europe n°1, 1960, 00:02:54, Archives Europe 1, Paris.

throughout the entire corpus, therefore minimising the impact of Klein's quote.

Giving the impression that discussions and comments are unscripted is one of the applications of liveliness. However, genuine unscripted events are also developing this experience of shared listening at the core of the section. One example is found in an amateur recording of evening programmes of RTL in 1967, in which the host presenting *Retour en musique* appears confused in two instances. First, before playing a contemporary cover of one of Bach's harpsichord concertos, he hesitates, stutters, and pauses, as he cannot recall the name of an artist. After the record, he clearly goes off-script: 'Well, as it would be a shame to play my jingle, *euh*, very rhythmic, almost thunderous, after this wonderful music, I am going to wish you goodnight, I..., I... [stutter] Anyway, it's only a "see you later" because we meet again at midnight, to cut the night short, from midnight to three in the morning. I hope that [pause] in my sweet company you [pause] you will have gotten home nicely, and in music'. Following these confused comments, he likely senses that time is running out and rushes to announce the journalist about to present the news. It is safe to assume that these hesitations and errors, found in the fabric of the host's radio speech, are genuinely unscripted, as they convey a sense of amateurism, and reinforce the idea of liveness. The decision not to play a jingle would usually not have an impact, unless, as it is the case here, the host clearly mentions it on air, stating a change in the script. To quote David Hendy, the rushed last words of the host to introduce the news reinforce liveness as radio's 'temporal rhythms'.[37] In this instance, Hendy would be referring to radio working on hourly, daily, and weekly cycles, and connecting with the rhythm of daily lives,[38] inherently incarnated here by the need to finish on the clock for the news. These references to radio's temporal rhythms are not easy to study within the audio corpus, as most of the sources were recorded by the broadcasters, therefore not including them as they are perceived as exterior to the programmes. The section below on sonic icons, however, allows a peak at these elements.

[37] David Hendy, *Radio in the Global Age* (London: Polity Press, 2000), p. 178.

[38] Media play an important role in shaping daily schedules, as pointed out by Gripsrud: 'broadcast media form a "normal" time scheme for "normal" lives, they structure each day and give each day of the week a special character'. Jostein Gripsrud, *Understanding Media Culture* (London: Arnold, 2002), p. 24.

Some amateur historical recordings, specifically two sources from Radio Luxembourg 208, are particularly insightful to understand the changes in the aforementioned cycles. In the first sound document,[39] a recording from September 1971, one can hear the end of the Dutch service, presented by Felix Meurders, and the beginning of the English service, presented by Paul Burnett. The Dutch disc-jockey appears to be listing, in Dutch, a series of records, with a fast pace and an enthusiastic voice, followed by Joan Baez's 'Here to you, Nicola and Bart', a sonic icon, marking the end of the Dutch programme. Meurders makes more comments in Dutch, mentioning the names of Dave Christian, Bob Stewart, and Paul Burnett.[40] He then switches to English and says: 'Thank you, ladies and gentlemen, for tuning in, switch you off, and turning in, and, hmm…', then going back to speaking Dutch, and then, once again in English: '…and to my English colleagues over there, in the Villa Louvigny, in Luxembourg. I really don't know who's coming out now [someone else laughs in the background] but I will find out. And, also, to my listeners in the Benelux, thanks very much, until next time, that will be on Sunday'. This is followed by a series of electronic sounds, before a male voice, Paul Burnett's, takes over: 'Oh, there you go, that's one of the, hmm, really crazy guys, from the Dutch service of Radio Luxembourg, from 6 O'clock till 7.30. That was Felix Meurders, thank you, Felix. My name is Paul Burnett, standing by for the English service of Radio Luxembourg'. Following Burnett signing on, a first record is played. In the second source,[41] the recording unfortunately does not capture the German disc-jockey signing off, although, it is possible to hear Tony Prince, from the English service, talking to his co-worker, Hans, and saying a few words in German and English to listeners: 'Guten Abend mein Freunds [sic] in Deutschland, this is the Tony Prince show, on the English [insists on the word] service of Radio Luxembourg'. These two unusual recordings are fascinating illustrations of radio's temporality, marking the changes in daily cycles. Yet, in these

[39] Felix Meurders, Paul Burnett, Radio Luxembourg 208, 2nd September 1971, 01:03:38, personal archives of Marius Zuiraitis.

[40] Dave Christian (1949–2010), Bob Stewart (1939–2019), and Paul Burnett (born in 1943) were all disc-jockeys working for the English service of Radio Luxembourg at the time.

[41] Tony Prince show, Radio Luxembourg 208, 23rd September 972, 00:59:43, personal archives of Marius Zuiraitis.

two occasions, the changes also include a change of language and audience,[42] something inherently different from other stations of the period. The discussions between two disc-jockeys also provide the sense of liveness of the station as it gives a taste of interactions within the studios of the Villa Louvigny.

As mentioned, liveness is an overall characteristic of radio, although its impact on listeners means it plays a strong role for commercial radio stations. A closer look at recordings of these stations reveals the richness of sonic elements playing into this listener engagement for a feeling of a shared radio experience. Liveness is a key aspect of media, particularly when accompanying extraordinary media events, such as those studied by Dayan and Katz.[43] Nevertheless, it also accompanies more 'mundane' moments of radio, fuelling this sense of shared experience. As shown previously in this chapter with the example of the interview with *The Shirelles*, during which the host is clearly lying about the liveness of the segment, one can see how important it was for commercial radio stations to maintain this sense of liveness.

6 Sonic Icons

Commercial radio stations were not the only ones using sonic icons and jingles; all stations did, but how sonic icons were used by Radio Luxembourg and Europe n°1 is the focus here. Jingles are sonic icons that were taken seriously by commercial radio stations. A key example of this being the French service of Radio Luxembourg, which, in 1964, designed a five-second jingle for the station that is still used today.[44] This jingle is composed of only a few notes of three brass instruments, trumpets and

[42] Although, the very moment of change between programmes could also be mixing languages for a short time, as illustrated in the two previous examples.

[43] Daniel Dayan & Elihu Katz, *Media Events: The Live Broadcasting of History* (Cambridge: Harvard University Press, 1992).

[44] The jingle was composed by Michel Legrand (1932–2019), a French music composer, known for writing the music scores for cinema and television. At the time, he composed for Radio Luxembourg, he was known for his work with directors from the French *Nouvelle Vague*. Among other examples, he composed the music for Agnès Varda's *Cléo de 5 à 7* in 1962, Jean-Luc Godard's *Une femme est une femme* in 1961 and *Bande à part* in 1964, and Jacques Demy's *Lola* in 1961 and *Les parapluies de Cherbourg* in 1964. It is safe to assume he was already a well-known and respected music composer by the time he collaborated with the radio station.

trombone, going in a canon. In 2009, Michel Legrand spoke on RTL about composing this jingle, remembering the station's request to have a short theme, to be used at all times, especially for the news. Legrand also commented on the process leading to its creation, including the addition of other brass instruments to make the theme slightly longer. He then noted the jingle became 'like a flag, like a sort of emblem',[45] a highly visual image for a sonic icon, revealing this idea that it clearly identifies the station to listeners. For its iconic programme *Salut les Copains*, Europe n°1 also appears to have gone to great lengths to create its jingles, as at least two famous singers of the time, Sylvie Vartan and Claude François, recorded short tunes for the station.[46] The jingle with Vartan is a short song, around fifteen seconds, with lyrics going: '" = "All your favourite slow songs / Salut les copains' on a slow pace, accompanied by bass guitar and a chorus of male voices. François' jingle seems inspired by a love song and the instruments usually associated with it, and lyrics going: 'And, here we go, it's SLC [abbreviation for *Salut les Copains*] / Starting again / We will laugh, we will sing / And have fun'. Both jingles echo the discography of the artists, reinforcing the association between artists, the station, and the show, which is mentioned in the lyrics. This association is particularly interesting as sonic icons such as jingles help identify a station and are a component of its image. As the aforementioned jingles included the artists' own voices—likely well known to the audience—they were integrating the artists' 'sound' into the station's.

In an amateur recording of *Salut les Copains* from 1966[47] consulted in Europe 1 archives, lasting slightly over an hour, thirteen different jingles can be counted, meaning that, on average, there was one jingle every five or six minutes, which is likely the most intensive use of these sonic icons among the programmes listened to. Even though all of the 13 jingles are different, they are bound by a few common characteristics: they only last a few seconds, find their roots in popular music genres (rock, yéyé, chanson, and jazz), and mention, without fail, the show, either through its full name (Salut les Copains) or by the abbreviation (S.L.C.) or even

[45] Stéphane Carpentier, 'Mort de Michel Legrand: comment il a composé le jingle de RTL', 26th January 2019, RTL.fr, [https://www.rtl.fr/culture/musique/mort-de-michel-legrand-comment-il-a-compose-le-jingle-de-rtl-7796345958] (last consulted 05/03/2023).

[46] However, the exact date is not known, nor how often they were used on the show.

[47] Salut les copains, 1966, Europe n°1, 01:08:00, Archives Europe 1, Paris.

both. As mentioned, jingles differentiate and serve specific functions in the structure of the programme, as some are used to indicate the arrival of a guest[48] or the beginning of a new section of the show.[49] Others give the postal address to write to the station (such as the postal box 150, mentioned previously), while others remind listeners of the show's timeslot (from 5 to 7 pm). The name of the host, Daniel, is also implemented in some of the jingles, joining his image with that of the show's. Interestingly, in the same year as this recording, Rosko started presenting his show *Minimax* on RTL, and his name was also woven into the jingle, as well as the address of the station (22, rue Bayard, in Paris), suggesting similar practices between the two stations.

The English service of Radio Luxembourg was also making use of jingles, although they appear to have been used mostly to identify the station and its frequency (208 m), whereas Europe n°1 and RTL refer to their postal address on a regular basis. The medium-wave frequency of the English service appears to be intersectly linked to its image and its identity, as illustrated in a recording of Mark Wesley's show, when he takes over Romeo, his Dutch counterpart, from 1971. Wesley, signing in, says: 'we're broadcasting from the facilities of the Villa Louvigny's studios in the heart of Luxembourg, through the powerful transmitter at Marnach on 208 metres, that's 1439 kHz on the medium waves. This is of course the English service of Fabulous Radio Luxembourg!'. This is followed by a jingle for Radio Luxembourg. It is but one occasion in which disc-jockeys mention the wavelength and the transmitter, and it is particularly interesting to see how this association between technical aspects and station's identity anchors the wavelength as part of the sonic icons. The use of sonic icons in commercial radio stations does not stop to jingles, as illustrated on Europe n°1's *Vous êtes formidable*, broadcast on Christmas Eve 1961,[50] in which the host presents a charity initiative for paralysed people. Inspired by Lyon's annual festival of light in December, during which inhabitants would light a candle by a window

48 In this instance, it is young women participating in a beauty pageant organised by another magazine owned by Filipacchi, *Mademoiselle Age Tendre*, who are invited for short interviews on air.

49 One example, taken from the recording, is a quick game, in which listeners have to guess a record based on a riddle read on air by the host.

50 Vous êtes formidable, 24th December 1961, Europe n°1, 00:13:05, Europe 1 Archives, Paris.

to celebrate the Virgin Mary. On this occasion, the station had the idea to extend this tradition all over France to show the support for the charity all over the country. The programme includes recordings of church bells in various French cities, used as sonic illustrations of their support for the initiative. These recordings include, among others, two churches in Lyon, the cathedrals of Strasbourg and Reims, church bells in Amiens, Albi, etc. Each extract sounds different, as a way to represent the singularity of church bells around France, each being an element of the wider soundscape. This use of bells as sonic icons for French cities echoes Alain Corbin's famous work on French soundscape and its church bells in *Les Cloches de la terre*.[51]

When analysing the sound of commercial radio stations by looking at the wider corpus of audio sources, this chapter has made apparent a series of sonic features specific to Europe n°1 and the French and English services of Radio Luxembourg. While characteristics such as liveness and interactivity are present in other radio stations, other features, such as playfulness and commercialism, are specific to commercial radio. It is, however, the combination of all the features that constitutes the fabric of the 'sound' of commercial radio stations. As such, this 'sound' plays an essential role in differentiating commercial stations from public broadcasters. In addition to the five aforementioned features of commercial radio, another element specific to broadcasting should be added. The concept of flow is crucial to the understanding of radio overall, and should therefore be mentioned. The notion of flow was first coined by Raymond Williams, who used it to describe his experience of American television as 'a single irresponsible flow of images and feelings'.[52] By doing so, he presented the flow of television as a stream of segments and advertisements all linked together. While there is some truth to this statement, Gripsrud has later shown that the flow could be applied to other media,

[51] In his iconic work, Corbin analysed the importance of bells in France and their network around the country in the nineteenth century. Through this research, he showed their importance in French history, and how they were connected to many social and political changes. Last, his research focus was on hearing, and put the sounds of bells at the core of his work. See Alain Corbin, *Les Cloches de la terre: paysage sonore et culture sensible dans les campagnes au XIXème siècle* (Paris: Albin Michel, 1994).

[52] Raymond Williams, *Television: Technology and Cultural Form* (New-York: Schocken Books, 1975), p. 91.

such as radio, and not just to television.[53] Moreover, the image of the flow carries a negative connotation, something that overflows audiences, a statement that should be nuanced. John Ellis did so—for television—when he made a distinction between the notions of gaze and glance. The gaze is associated with cinema and a more precise attention of audiences to the medium. The glance, mainly associated with television, but which may also be used in the context of radio, offers a more nuanced and distant relationship to the medium.[54] Returning to the image of the flow, Gripsrud explains that the concept makes sense to describe audiovisual programmes that follow one another and are interspersed with other segments in a coherent fashion. He insists that there should be more distance with the place of audiences. Audiences, such as listeners of commercial radio stations, do not have to be in the middle of the flow when consuming such media. They can indeed stand at a distance, and observe this flow.[55]

7 The Materiality of the Commercial Radio Experience

While radio might be a medium relying on airwaves and the experience of radio might appear as an immaterial one, it is actually strongly anchored in materiality, and its sound plays a part in this anchoring. This presence of materiality is understood as audio references, whether speech or sound, that point to the physical production and reception of radio waves. The references can be through objects and devices, *i.e.* transistors and radio sets, used in this regard, but also the spaces of production and reception, such as studios and microphones. This presence is separated into two categories, based on the nature of the sounds. On one hand, there are diegetic elements that might call the listener to come closer to the radio set, to turn the volume up, and inform us about the recording conditions, such as technical issues and noises, whether they are voluntary parts of the radio programmes or not. On the other hand, some recordings offer

[53] Gripsrud, *Understanding Media Culture*, p. 213.

[54] John Ellis, *Visible Fictions. Cinema, Television, Video* (London: Routledge, 1982), p. 116.

[55] Gripsrud, *Understanding Media Culture*, p. 213.

non-diegetic elements[56] that can enrich the understanding of the sounds surrounding and modulating programmes of commercial radio stations (*e.g.* interferences, background sounds). Moreover, it informs us of other sounds that would have been heard alongside the radio programmes. Depending on the nature of their recording, some sources can also help understand, to some extent, the ways in which radio sets might have modulated the original programmes. These elements are external to the radio programmes themselves but a part of the radio experience nonetheless. Both diegetic and non-diegetic categories are useful to further the understanding of the very experience of listening.

Non-Diegetic Sound Elements. Some of the sound sources consulted for this research accidentally informed on the listening and recording conditions at the time. These sources were not recorded by the broadcasting entities, which is how they came to offer these extra sonic indications. While these layers lower the sound quality of the original recordings, they are actually beneficial to understand these questions of radio materiality. Available in the archives of the INAthèque, a series of recordings of evening radio news from commercial radio stations are particularly interesting for this listening experience. There is little to no information about these documents apart from the recording dates. However, they are fascinating audio sources, likely made by a recording device put in front of a radio set. In one of the recordings, from 1963,[57] the evening news can clearly be heard, the journalist's voice is perfectly audible, but light static can be heard throughout the whole document, in addition to a second male voice in the background, separate from the original broadcast. This second voice, both distant and muffled, also seems to be speaking French, but it is not possible to discern actual words. The tone and inflection of the second voice indicate it is another journalist reading the news, but on another station, which is quite logical considering news was read at the same time, on the hour, on most stations.

[56] As mentioned, a non-diegetic sound is a sound 'whose apparent source is not the space-time of the scene depicted'. Michel Chion, '100 concepts pour penser et décrire le cinéma sonore', bilingual version, October 2012. [http://michelchion.com/] (last accessed 27 March 2023).

[57] Journal parlé d'Europe n°1, Europe n°1, 31st July 1963, 00:22:03, INAthèque, Paris.

Another source,[58] made of recordings of radio news on Europe n°1 at three different times of the day, offers a similar experience in its third extract, in which a second voice from another station can be heard in the background. These sound elements belonging to another radio station heard in the background appear in other recordings. In one of them, the secondary voice that can be heard does not speak French.[59] The author was unable to identify the language used, although the voice was clearly male and likely iterating the news as well, as suggested by its speech and rhythm.

On other occasions, interference coming from other stations was not from a voice, but rather from music. In a source comprised of recordings of radio news from Radio Luxembourg and Europe n°1,[60] a joyful song can be heard in the background, alongside Europe n°1's journalist telling the odd story of a waiter—*un garçon de café*—who accidentally got lost in the sewers for roughly three hours until someone in the streets heard him cry for help and then saw him waving his hand through a manhole. This interference is not simply an extra layer; it is a layer that, due to the juxtaposition between the story and the happy song, creates an unplanned tension that turns the poor waiter's misfortune into an accidentally funny anecdote. This shows that these intradiegetic elements can have an effect that is more powerful than a simple disturbance within the content of radio programmes.

In a recording of the news of Europe n°1 from February 1959,[61] a percussion rhythm, followed by some guitar notes, can be heard, discreetly, alongside the voice of the journalist reading the news. Moreover, this sound document includes strong static, meaning that three sound layers can be identified: the main one, or, the original voice of Europe n°1's journalist presenting the news of the day, the sound layer coming from the other radio station, and the interference covering the whole recording. It is relevant to mention that the last two layers are

[58] Journaux parlés d'Europe n°1, Europe n°1, 27th August 1962, 00:23:13, INAthèque, Paris.

[59] Journaux parlés d'Europe n°1 et de Radio Luxembourg, Europe n°1 & Radio Luxembourg, 2nd September 1962, 01:12:09, INAthèque, Paris.

[60] Journaux parlés d'Europe n°1 et de Radio Luxembourg, Europe n°1 & Radio Luxembourg, 10th July 1963, 01:07:07, INAthèque, Paris.

[61] Journal parlé d'Europe n°1, Europe n°1, 18th February 1959, 00:09:24, INAthèque, Paris.

linked to the listening device, and would not be present in recordings made by broadcasters. These amateur recordings do, however, reflect some aspects of what may have actually been heard by listeners.

This is important as it creates a dichotomy between the consulted audio sources; there are the 'pristine' recordings, from the production side; and there are amateur sources, which might include interferences. Although these interferences may actually reflect, in some way, how the 'sound of commercial radio' was actually heard by listeners, in a more precise manner than with 'pristine' recordings. As already mentioned, it is not possible to fully capture this sound. However, including these sonic elements and interferences is a step towards shedding further light on the differences between production and reception of commercial radio.

One aspect of radio that has not been mentioned is the listeners' ability to change stations. Yet another group of sources found at the INAthèque in Paris offers insight into the act of listeners changing stations. In a recording of the evening news,[62] it is possible to hear the listener changing the station from Radio Luxembourg to Europe n°1, the voice of the journalist stopping abruptly, followed by static and noise, until another voice is heard. Both stations are identified thanks to sonic icons, their jingles, as well as adverts that mention the name of the station, which are important since the news follows similar formats and the content is also comparable. Such a recording highlights the transnational mind map of the radio dial[63] and makes an audio reality; the listener—in an unknown location—tuning in to stations whose transmitters are in Luxembourg and Germany, this change in airwaves being materialised into a series of sounds.

Diegetic Sound Elements. The other aspect of the materiality present throughout the audio corpus used in this research consists of sound elements within the programmes, in what is said and in what is heard on air. These elements are clear references to the materiality of radio experience and connect the programmes to their conditions of production and/or reception, creating bridges between the soundscape and the radio experience.

[62] Journaux parlés d'Europe n°1 et de Radio Luxembourg, Europe n°1 & Radio Luxembourg, 28th August 1962 [duration unknown] INAthèque, Paris.

[63] Andreas Fickers, "Visibly Audible. The Radio Dial as Mediating Interface", in T. Pinch & K. Bijsterveld (eds.), *Oxford Handbook of Sound Studies* (Oxford: Oxford University Press, 2012), pp. 411–439.

One mention of the radio experience can be found, for example, in the evening news on Europe n°1 in 1963,[64] during which the following advert can be heard: 'This commercial break is devoted to a little experiment. Listen carefully. Your transistor, yes the transistor with which you are currently listening to Europe n°1. Well, equip it with special transistor Mazda batteries. Right away, you will hear the difference. With the special transistor Mazda battery, Mayer audition, enhanced power. And also, you will make another discovery with a special transistor Mazda battery: the listening length is extended by thirty per cent'. This transcribed advert is an example of occurrences of radio programmes directly addressing the listening devices. These devices are an essential cog in the radio listening experience and in the *dispositif* of commercial radio overall. References to the devices, and how they influence the sound quality, bring the material reality of radio into the programme. In another recording, the host addresses listeners at the end of a live performance by French-Armenian singer Charles Aznavour and explains they should ponder on their luck: they only have to turn a button, and thanks to the miracle of radio, Charles Aznavour is in their home.[65] Here again, this comment conjures the materiality of radio listening and the physical action of turning on the radio set. It also plays with imaginary references linked to radio, this 'miracle' and the presence of the artist in the house.

In *Pour ceux qui aiment le jazz* broadcast in 1955, what is happening in the studio and the way radio is being made is quite present through various sounds and comments by the hosts. First, Daniel Filipacchi and Frank Ténot are discussing a record in a literal sense[66]:

> Frank Ténot: Good evening, Daniel, how are you?
> Daniel Filipacchi: Ah, very well, very well. I see a record over there...
> F.T.: Wait, wait. Don't take it. Leave it, leave it.
> D.F.: [laughs] Why wouldn't I take it?
> F.T.: Because we're going to play it now
> D.F.: Alright, but let me look at the picture. Who's that fat guy?
> F.T.: It's a record by Clark Terry.
> D.F.: Oh, it's Clark Terry?

[64] Journal parlé d'Europe n°1, Europe n°1, 31st July 1963, 00:22:03, INAthèque, Paris.

[65] Musicorama, Europe n°1, 9th July 1957, 01:01:04, Archives Europe 1, Paris.

[66] Pour ceux qui aiment le jazz, Europe n°1, 1955, 00:20:30, Archives Europe n°1, Paris.

Second, multiple different sounds from the studios can be heard. Objects, perhaps glasses and cups, are put down on a table, a chair squeaks, papers are shuffled, clothes are removed, etc. Fifteen minutes into the programme, one of the hosts explains how listeners sometimes complain that their live orchestra is not truly live and that it is rigged. To prove such speculation wrong, the host addresses the musicians individually and asks them to play a few notes to prove to the audience that musicians are currently present in the studios and that their performance is in fact live.

Francis Blanche explains he had been in the studio for a moment already: 'It has been at least five minutes since I entered, quietly, without being noticed, and that I'm here, peacefully, in my corner, I listen, I look, I think, I ponder…'[67] The same host, in another show the following year,[68] sends his co-host to a corner[69] of the studio for a time-out as punishment. The co-host can be heard walking a few steps, and his voice becomes more distant on the tape, giving the audience a sonic perception of the space of the studio. Blanche asks him then to put his hands on his head as a follow-up punishment, but the co-host refuses, explaining that he might drop the sheet with his text. He therefore must stand on one foot, which causes him to lose balance and fall to the ground. Through the use of both speech and sound, this short interaction offers a vivid depiction of the studio's space, a location that usually disappears on radio. The interaction also gives a clue about the way radio is done, through this mention of the sheet used, something that is usually not talked about to preserve an unrehearsed image of radio. The sonic presence of the studio is also particularly present in a programme called *Dans le vent*, hosted by Michel Cogoni and broadcast in December 1963 on Europe n°1. In it, he receives the French singer and actress Dalida, however, the show is marked by several disruptions. First, five minutes into the programme, one can hear multiple footsteps as a background sound, as if a group of people were bringing something heavy into the studio. The host therefore is heard (his voice is heard from a distance and one can hear his body

[67] Le Sourire de Francis Blanche, Europe n°1, 1960, 00:02:54, Archives Europe n°1, Paris.

[68] Midi à 14 heures, Europe n°1, 26th March 1961, 00:44:28, Archives Europe n°1, Paris.

[69] In French, the expression *aller au coin* means, for children, to be put on time-out, and sent, literally, in the corner of a room.

moving) turns to these men and addresses them with a mean comment. Second, he later explains to his guest: 'If you hear sounds from time to time, they are sounds of a duplex installation, with… Erm, because we're planning stuff', which is illustrated by numerous barely audible background conversations with unidentified people. Third, Cogoni is interrupted in the middle of a question by a ringtone. He then picks up the phone and angrily answers the unidentified caller: 'Oh, come on, I'm not available! […] Yes, what is it? You want to take over my programme? […] What is it? […] Yes, well, *voilà*'. he hangs up and resumes his conversation with Dalida as if nothing happened. Phone-ins are a regular practice in radio, however, usually both parties can be heard by listeners. This specific interruption is highly unusual in the way that it informs on the materiality of radio production. Fourth, the host directs his guest who has to perform a song, and gives her many directions that inform on the space of the studio: 'Take the sheet'; 'Go there'; 'Look over here'; or 'Go sing', especially as they are accompanied by her footsteps and her voice being heard in various locations in the space.

The materiality of the radio experience is not solely located in the studio, but also where radio is listened to, and, in this regard, the comments of the female host of *Le Mot de Passe* on Radio Luxembourg in 1959 are highly informative: 'Good evening, dear listeners, here I am, like every Sunday, loyal to our *rendez-vous*, I hope you are numerous, and that you, *Madame*, who are maybe making the Sunday cake, you, *Mademoiselle*, who might be ironing your dress for this afternoon's dance, and, you, *Monsieur*, who might tinker in a corner of the house, well, will lend me a lenient ear and…'[70] These comments are insightful regarding the perception of listeners but also regarding listening practices, as the host suggests the programme is listened to while performing house chores (described in a rather stereotypically gendered way), implicating the assumed contextualisation of listening.

By studying audio sources as a whole, including sounds which do not belong to the programmes broadcast, this section fights the impression of radio as an immaterial medium. The analysis of both diegetic and non-diegetic sound elements indeed anchor the sources in the materiality of their context of reception and production. While an impromptu

[70] Le Mot de Passe, Radio Luxembourg, 1959, 00:15:00, INAthèque, Paris.

and disruptive phone call on-air points to the ways radio is done, recordings where another station can be heard in the background shed some light on what listeners actually may have heard at time. In both instances, what implicitly appear are various constitutive elements of the *dispositif* of commercial radio; *i.e.* the space of the studio, the accidental phone call, and the changing sound quality of the stations, which all belong to the heterogeneous ensemble studied throughout the book.

References

Aalbers, Jasper. "Echoes of the City: Staging the Urban Soundscape in Fiction Film". PhD Doctoral Thesis, Maastricht University, 2013.

Antoine, Frédéric. *Analyser La Radio. Méthodes et mises en pratique*. Louvain-la-Neuve: De Boeck, 2016.

Bernard, Luc. *Europe 1: La Grande histoire dans une grande radio*. Paris: Centurion, 1990.

Bijsterveld, Karin (ed.). *Soundscape of the Urban Past*. Bielefeld: Transcript, 2013.

Carpenter, Marie. *La Bataille Des Télécoms. Vers Une France Numérique*. Paris: Economica, 2011.

Carpentier, Stéphane. "Mort De Michel Legrand : comment Il a Composé Le Jingle De RTL". RTL.fr, January 26th, 2019. [https://www.rtl.fr/culture/musique/mort-de-michel-legrand-comment-il-a-compose-le-jingle-de-rtl-7796345958]

Chignell, Hugh. *Key Concepts in Radio Studies*. London: Sage, 2009.

Chion, Michel. "100 Concepts Pour Penser et Décrire Le Cinéma Sonore". Online, October 2012. [http://michelchion.com/]

Clark, Ray. *Radio Caroline: The True Story of the Boat that Rocked*. New York: The History Press, 2014.

Corbin, Alain. *Les Cloches de la terre : paysage sonore et culture sensible dans les campagnes au XIXe siècle*. Paris: Flammarion, 1994.

Dayan, Daniel and Elihu Katz. *Media Events: The Live Broadcasting of History*. Cambridge: Harvard University Press, 1992.

Ellis, John. *Visible Fictions. Cinema, Television, Video*. London: Routledge, 1982.

Fickers, Andreas. "Visibly Audible: the radio dial as mediating interface". In *The Oxford Handbook of Sound Studies*, edited by Trevor Pinch and Karin Bijsterveld, 411–439. Oxford: Oxford University Press, 2012.

Glévarec, Hervé. "Le reportage radiophonique des 'événements' : Mai 68 a-t-il inventé 'l'effet de réel' en radio ?". In *Images et sons de Mai 68 (1968–2008)*, edited by Christian Delporte, Denis Maréchal, Caroline Moine and Isabelle Veyrat-Masson, 43–53. Paris: Nouveau Monde éditions, 2011.

Gripsrud, Jostein. *Understanding Media Culture*. London: Arnold, 2002.

Griset, Pascal. "Le développement du téléphone en France depuis les années 50. Politique de recherche et recherche d'une politique". *Vingtième siècle, revue d'histoire* 24 (1989): 41–54.

Hendy, David. *Radio in the Global Age*. Cambridge: Polity Press, 2000.

Legay, Richard. "RTL & Europe n°1 as central actors. The importance of mobility for commercial radio stations during the Parisian events of Mai 68". *Rundfunk & Geschichte* 3:4 (2018): 41–50.

Legay, Richard. "The Role of Commercial Radio Stations in the Media Vacuum of Mai 68 in Paris". *VIEW Journal of European Television History & Culture* 6:12 (2017): 1–11.

Rikitianskaia, Maria. "A Transnational Approach to Radio Amateurism in the 1910s". In *Transnationalizing Radio Research: New Approaches to an Old Medium*, edited by Golo Föllmer and Alexander Badenoch, 133–140. Bielefeld: Transcript.

Russo, Alexander. *Points on the Dial. Golden Age Radio beyond the Networks.* Durham: Duke University Press, 2010.

Scannell, Paddy. *Media and Communication*. London: Sage, 2007.

Schmitt, Marie-Paule. "Les Jeux sur les ondes de la Libération aux années soixante-dix : grandeur et décadence d'un genre radiophonique". *Sociétés & Représentations* 24:2 (2007): 353–369.

Schmitt, Marie-Paule. "Les Jeux radiophoniques, en France, dans les années soixante". *Sociétés & Représentations* 37:1 (2014): 219–228.

Spinelli, Martin. "Masters of Sacred Ceremonies", In *Broadcasting Modernism*, edited by Cohen, Debra Ray, Michael Coyle and Jane Lewty, 68–88. Gainesville: University Press of Florida, 2009.

Stamm, Michael. *Sound Business: Newspapers, Radio, and the Politics of New Media*, 2nd ed. Philadelphia: University of Pennsylvania Press, 2016.

Tolson, Andrew. *Media Talk: Spoken Discourse on TV and Radio.* Edinburgh: Edinburgh University Press, 2006.

Wagner, Hans-Ulrich. "Sounds like Hamburg... Hamburg-Klänge Und Mediatisierte Raumkonstruktionen in Der Frühzeit Des Radios". In *Radio Und Identitätspolitiken: Kulturwissenschaftliche Perspektiven*, edited by Johannes Müske, Golo Föllmer, Thomas Hengartner and Walter Leimgruber, 55–72. Bielefeld: Transcript, 2019.

Williams, Raymond. *Television: Technology and Cultural Form.* New York: Schocken Books, 1975.

CHAPTER 7

Sharing the Airwaves: A Collaboration Between RTL and Europe n°1

- Je crois que maintenant il va falloir dire très, très vite au revoir aux auditeurs d'Europe Luxem-1…

- …et de Radio numéro-bourg ![1]

- Jean Yanne & Francis Blanche, in the evening of 8th May 1968, on RTL & Europe n°1.

This book supports the idea that the main commercial radio stations can be, and should be, studied as a whole and that there is an interest in gathering them under one banner; to use the French expression, the banner of the *postes périphériques*.[2] This is why this chapter is an analysis of a specific source, a common show between RTL and Europe n°1, broadcast in May 1968,[3] to show that the closeness mentioned could materialise, and offer an example of a moment in which the sounds of both stations would blend into one for a little while. The common show

[1] 'I believe that we now have to very, very quickly say goodbye to the listeners of *Europe Luxem-1*…

…and of *Radio numéro-bourg*' (parodic names given to the two stations by mixing their names in French).

[2] A French expression referring to commercial radio stations whose transmitters were installed outside the French border.

[3] Common radio programme, 8th May 1968, RTL & Europe n°1, 02:31:58, personal archives of Roland Biesen.

R. Legay, *The Cultural Histories of Radio Luxembourg and Europe n°1*, https://doi.org/10.1007/978-3-031-46250-4_7

by RTL and Europe n°1, as a source, is a tremendously rich and unusual—if not unique—recording, which is why it deserves its own chapter. As a source, its uniqueness resides in the fact that it is a radio programme made in conjunction with two French-speaking commercial stations, RTL and Europe n°1, as well as because it was an experiment with stereophony. A longer analysis of the programme's content reveals the ways the stations shaped, and in some ways, directed, the listening experience of their audience. It also reveals the complex dynamics operating between the stations to shape their own images, either individually or as a single entity ('*les périphériques*') in the wider media landscape. Other aspects, such as the irreverence and the playfulness of the hosts, as well as the technical *coup de maître*, are addressed in this chapter, as they play crucial roles throughout the recording.

This radio programme was recorded at the time of its broadcast, and later digitised, by a former Europe n°1 technician, Mr. Roland Biesen, who provided the author with a digital copy. The programme, in French, was made and broadcast live in common by two commercial radio stations, RTL and Europe n°1, on the evening of Wednesday 8 May 1968. It is a programme of light entertainment, lasting around two-and-a-half hours, made-up of a succession of short sketches, records, advertisements, discussions, and jokes, between the two hosts, Jean Yanne[4] (for RTL) and Francis Blanche[5] (for Europe n°1). Both hosts were quite famous at the end of the 1960s for their involvement in several aspects of the

[4] Jean Yanne (1933–2003) was a French humorist, mostly known for his work as an actor. He played in an extensive list of movies since the 1950s, although he was also known for his work as a movie director, his work on radio and as a writer, including a graphic novel, *Les Dossiers du B.I.D.E.*, at the end of the 1960s. He started his radio career on the airwaves of Europe n°1, before moving to RTL after his dismissal. He was dismissed again from RTL at the end of 1968, and therefore moved to France Inter. He came back to RTL in 1977 as a member of the long-lasting show, *Les Grosses Têtes*. His successive dismissals found their roots in tensions with the stations due to its irreverent sense of humour, at times misogynist, and often highly critical of authority, institutions and publicity. Denis Maréchal, *RTL. Histoire d'une radio populaire. De Radio Luxembourg à RTL.fr* (Paris: Nouveau Monde Éditions, 2010), pp. 364-367.

[5] Francis Blanche (1921–1974) was a French humorist and actor, who started his career in theatre, although he might have been best known for his songwriting and his collaboration with Pierre Dac (1893–1975) on many radio programmes, including *Signé, Furax* on Europe n°1, a highly popular daily humoristic drama. Blanche also wrote humoristic lyrics for a large number of songs and played in dozens of movies from the 1940s until his death.

entertainment industry, whether it was as actors (theatre and movies), songwriters, singers, or on air. The joint programme could be listened to on both stations simultaneously and the hosts encouraged the listeners to use two transistor sets, each tuned in to a different station, in order to enjoy the full stereophonic experience. It was recorded in a similar fashion at the time, following the hosts' instructions. If listened to with proper audio equipment, the digitised recording does convincingly recreate the experience of listening to two sets, one on each side.

The context of the broadcast of this show is rather interesting as it is at the beginning of May 1968, a period of massive social unrest throughout France, especially in Paris during the early days. This unrest that was affecting the French capital, where the two stations had their studios,[6] is only briefly mentioned during the news. Both stations were involved in the coverage of the riots and protests that occurred during the Parisian events,[7] however, this show is a reminder of the separation between information and entertainment in radio stations schedules.

The idea of a 'media vacuum' during the events of May 1968 was used to explain the leading position of commercial radio stations in the French media landscape, in regards to the news, as a reaction to the near silence of the public broadcasting agency.[8] This source does not question this idea, but it illustrates the continuity of entertainment programming during this time of social and political unrest, alongside the key role of media coverage. In many ways, *Mai 68* can be seen as a 'media event', one that can mask, by its very nature, other broadcasting elements of the month of May 1968 that do not belong to this event. Moreover, the two radio stations were, in 1968, neck and neck when it came to the audience—however, still behind *France Inter*, part of the public broadcaster—turning this collaboration into an even more unexpected moment.

[6] RTL's studios were located in Rue Bayard, while Europe n°1's were Rue François 1er, only a few hundred metres away from each other, in the 8th *arrondissement* of Paris.

[7] Richard Legay, 'RTL & Europe n°1 as central actors. The Importance of Mobility for Commercial Radio Stations during the Parisian Events of Mai 68', *Rundfunk & Geschichte* (3:4, 2018), pp. 41-51.

[8] Richard Legay, 'The Role of Commercial Radio Stations in the Media Vacuum of *Mai 68* in Paris', *VIEW Journal of European Television History and Culture* (6-12, 2018), pp. 1-11.

The dramatisation of the image of the stations and their staff throughout the various sketches informs on the specificities of each station's image, while the overall irreverence characterising the hosts' performance points at shared characteristics specific to *les postes périphériques*. Moreover, the programme questions the listening experience and a non-staged potential cooperation beyond this specific show. Individual sketches throughout the joint programme open and close on various sonic icons like a siren or a gong, clearly indicating that the various parts of the programme are built around the central narrative of the discussion between Blanche and Yanne. Based on the differences between dialogues and sketches, it seems quite likely that the former was live. While the sketches have sound effects and illustrations (*i.e.* electronic tools and sounds of sewers) and an audio 'cleanliness' associated with pre-recorded programmes (although this section does not argue that this impacts the liveness of the show), the hosts are heard, at times, laughing and hesitating.

1 Shaping the Listening Experience

The recording starts on a series of comments by two unidentified men guiding listeners to place themselves in the best possible way before the programme. These comments, transcribed and translated[9] below, are noteworthy because the voices direct their audience on how to behave and where to sit, setting the stage for their listening experience. The first is a calm male voice, which—over a light background piano music, which is barely audible, similar to those heard in lifts and waiting rooms—explains: 'If we are here, now, it is to allow you[10] to make some technical adjustments, because, if you have your two receivers, you should hear the same modulation, you should hear us in the same way, on Radio-Télé-Luxembourg, left receiver, on Europe n°1, right receiver'. A second male voice, also calm, though slightly livelier than the first, takes over, as if interjecting in the middle of the explanation to add the following comments: 'Yes, well, I would like to clarify, right away, that if—unfortunately—you only have one receiver, stay tuned, because you will be able

[9] Translation made by the author.

[10] In French, they are using the word '*vous*', which can be used to address, formally, one person, or a group. It seems more likely to be the formal address to a single listener.

to follow our programme in perfectly acceptable conditions. And, if you have the chance, well, well, to have two receivers, place one on the left, as in a V-shape, if you wish, the other one on the right, as in a V-shape, and adjust the power properly. You, being in the middle, so the three—the two receivers and you—make a triangle, so you can hear **in the same way** [insists on the term] with the **same** [insists on the term] power, the two receivers. You need to have, currently, the impression that our voices are **in the middle** [insists on the term], and not that they are on your right or on your left'. The first voice then resumes the explanations: 'And to allow you the final adjustment, because, in the end, we have thirty seconds of extra time, we are now announcing a certain number of records that you will hear during this programme, and that will **not** [insists on the term] be announced during this programme, and we thank the artists who have given us the opportunity to play, as an exclusive, the last recordings which were made no later than this morning'. He then proceeds to list a series of French records (*i.e.* 'Cent fois ma vie' by Mireille Mathieu and 'La Bicyclette' by Yves Montand), before the second voice interrupts him to list some records of *variété étrangère*, which can be understood as non-francophone mainstream and popular music, that is also to be played (i.e. 'Michelle' by Andy Williams—originally by The Beatles—and 'Good Day Sunshine' by The Beatles). One of the voices then starts a countdown, from five to zero, concluded by a 'back to the stereo studios', which is quickly covered by a short, less-than-twenty-seconds, instrumental tune, performed by many brass instruments, audibly indicating the beginning of the common programme.

The pre-programme section described above is a particularly fascinating piece of radio. The two men on air give a series of instructions to their listeners, telling them how to position themselves and their receivers in preparation for the show. The hosts' description of an ideal listening space, composed as a triangle, with two receivers and a listener, is unique in the history of commercial radio stations. While references to listening devices exist in other programmes, they are always made according to the typical way radio listening is presented: 'one listener, one receiver, one station'. This programme, however, breaks this traditional construct by adding another radio station, and another receiver, disrupting the usual listening experience. The suggested way to listen to the common radio programme is unusual and requires preparation, which echoes the tinkering and 'playfulness' of early radio amateurs.[11] The suggestion of an ideal way to listen to the show means that listeners of both stations were sharing the same listening experience, or, at least, were encouraged to. In the particular case of the joint programme, the idea of an imagined community of listeners of commercial radio stations is pushed further. Not only do they take part in a 'ceremony replicated by thousands',[12] tuning in at the same time,[13] which already reinforces the power of radio in regards to imagined communities, but it also brings together the listeners of both stations under the same group. It is necessary, however, to nuance the impact of the show as nothing is known about the number of listeners. On one hand, little has been found about this joint programme, even in grey literature. On the other hand, it was broadcast in the middle of a specific time in France, something that could have occulted recollection of the joint programme in memories.

2 Setting the Tone

The programme itself begins, after a quick countdown, with some instrumental music, led by brass instruments (trumpets), clearly marking the official beginning of the programme in a serious, grandiose manner. The two hosts, Jean Yanne and Francis Blanche, greet each other, before exchanging a few comments about this joint show, celebrating, with humour, the occasion:

Jean Yanne: Well, for years, we have to admit it, the Radio War was not an empty word.

Francis Blanche: Indeed, the sky was heavy with lightning and clouds, until one day [he insists on the word day], elegiac voice [he pretends to read stage directions], until one day [sound of a bird flapping its wings],

[11] Maria Rikitianskaia, 'A Transnational Approach to Radio Amateurism in the 1910s', in G. Föllmer & A. Badenoch (eds.), *Transnationalizing Radio Research: New Approaches to an Old Medium* (Bielefeld: Transcript), pp. 133-140.

[12] Benedict Anderson, *Imagined Communities: Reflections on the Origin and Spread of Nationalism*, 3rd ed. (London: Verso, 2006), p.35.

[13] Michele Hilmes, 'Radio and the imagined community', in Jonathan Sterne (ed.), *The Sound Studies Reader*, (Abingdon-on-Thames: Routledge, 2012), pp. 351-362.

the dove of peace sang [unpleasant human-made bird call, moving from the left ear to the right].

J.Y.: Yes, that is not...that is not really a dove.

F.B.: No, but it is supposed to sound better in stereo.

J.Y.: Oh, right. Anyway, this symbolic dove's call shows you that now, peace exists, and completely, between the two stations. Francis even uses a neutral voice…

F.B.: Yes. [laughs]

J.Y.: Yeah, you have a… [laughs]

F.B.: Exactly.

J.Y: What have you been up to last night? [deeper tone, more casual voice, to give the impression he only speaks to his colleague]

F.B.: It is a great …[laughs]… it is a great event, nevertheless, this common evening RTL and Europe n°1.

J.Y.: Ah! It is the first time in radio history that it exists, and also the first time that it exists, in the field of stereophony, but it was not without troubles.

F.B.: Long and painful negotiations occurred, under our aegis, Jean and I...[slow pace, dramatic tone]'[14]

This short exchange transcribed above reveals the importance given by the two hosts to the unusual dimension of such a common show. They attempt to frame it as a historic moment, by relying on two main ideas; the technical prowess and the reconciliation narrative. First, the stereophony, presented as a novelty, is the key argument of the technical prowess argument, which benefited from the explanations of the pre-programme section. Second, the narrative of reconciliation is a recurring theme throughout the programme, although it is even clearer in the opening discussion quoted above. The references to war, either direct ('radio war') or in metaphor (through the theatrical references to clouds and lightning, for example), are key elements to the reconciliation narrative, that is overtly incarnated in the sound of the so-called dove, a well-known symbol of peace. It is clear that this narrative is taken with

[14] Translation by the author.

humour by the hosts, it nevertheless still plays with shared references to Western countries.

The technical prowess theme is further developed and staged in the programme within two consecutive sketches. In the first, the hosts recreate the alleged transplant operation with 'specialists of electronic surgery' that brought the two antennas together for the common show. The two characters are medical professionals—probably a surgeon and his assistant—over an operating table, which is audibly suggested by the electronic and buzzing sounds in the background that imitate the imagined soundscape of a hospital room, as well as the ways they address each other ('doctor' and 'professor'). The core of the humour at stake relies on each character asking for absurd made-up medical tools and a number of puns.[15] In the second sketch, the hosts stage negotiations between a technician from Luxembourg and a technician from the Saarland who tries to formulate an agreement to share the airwaves. The characters speak with heavy accents—allegedly from Luxembourg and Saarland[16]—and in a rather poor, broken, French, portraying the technicians as intellectually limited and greedy characters. This echoes the definition of comedy and humour offered by Chignell, in which he insists on the importance of using transgressive humour in radio programmes, which sometimes teeters on the verge of offensiveness,[17] which is something at stake in this depiction of the 'foreign' technicians. The use of foreign characters, who insist on their origins, is an interesting way to conjure the international characteristics of the stations and their collaboration, as well as the materiality of the technical prowess by referring to the transmitter sites, where broadcasting happens. This anchors the programme to material and geographical references that are often linked to the perceived freedom of the stations, which is in opposition to public broadcast.

[15] Unfortunately, such puns are almost impossible to translate into English, despite the author's best efforts.

[16] Based on the author's personal experience, accents in French from Luxembourg and the Saarland sound nothing like those in the recording.

[17] Hugh Chignell, *Key Concepts in Radio Studies* (London: Sage, 2009), p. 13.

3 Staging a Historic and International Encounter

The two hosts also explain, as part of the dramatisation around the joint programme, that they found a spot for a meeting of their directors, in the sewers under *Avenue Montaigne*, which is halfway between *Rue Bayard* and *Rue François 1er*, where the studios of RTL and Europe n°1 in Paris are located. These references to specific streets reinforce the local aspect of radio.[18] In this case, a Parisian aspect is reinforced, by setting the sketch in the French capital. The hosts then introduce an audio source into their programme, supposedly the radio report capturing the historical moment when the directors of both stations met to end their feud and make the common show happen. This short, fictional, and comedic sketch opens with the voice of Francis Blanche, who speaks softly so as not to disturb the scene, and with short sentences, imitating a journalistic style. He describes what is supposedly happening: the doors open and the two delegations walk towards each other, which is illustrated by the sounds of footsteps and creaking doors. Apart from his voice, one can hear sounds of water dripping, and rats squeaking, reinforcing the idea that the scene is happening in a sewer. The result is convincing, as it plays with the shared imagination of what sewers should 'sound like', by developing a rather irreverent background for the meeting of the stations' directors.

The recording moves on to male characters discussing. One voice is characterised by a heavy Southern French accent, while the others can only be described as 'silly voices' as they are distinguishable by highly nasal sounds and ridiculous tones. One of them, potentially a guard or policeman, leads the discussion and states he is there to check the validity of required official papers. He asks the other characters to produce a series of documents specific to the French administrative system.[19] The large quantity of documents asked, as well as their incongruity in such

[18] And these local aspects, known to many listeners, 'however mundane and ordinary, local content's familiarity and recognisability [have] value', Heather Norris Nicholson, 'Living on Location: Amateur Creativity and Negotiating a Sense of Place in Yorkshire', in I. Franklin, H. Chignell & K. Skoog (eds.), *Regional Aesthetics. Mapping UK Media Culture* (Basingstoke: Palgrave Macmillan, 2015), pp. 17-34.

[19] Among the documents, this character asks for '*le livret de famille*' (official booklet for the members of a French household), '*la carte grise*' (papers for a car or vehicle) and a weekly travel card for the Parisian '*métro*'.

a setting, shape the comical aspect of the scene, which can be described as rather irreverent towards the directors of both stations. This irreverence is particularly present when the two directors start making numerous elegiac comments about one another, while self-deprecating themselves. The characters' silly voices, combined with their pompous style and complex syntax—probably to illustrate the upper-class upbringing of a radio station's director—add an even more ridiculous dimension to these compliments that turn the overall dialogue into absurdity. One example that shows the compliment and the self-deprecation could be translated as such: 'I thank the admirable torch that you are, to agree to waste its lights to illuminate the dark and slimy worm that I am'.[20] The escalation of these compliments reaches a point where the two characters are on the verge of fighting over who is the most incompetent, before making up and kissing each other on the cheek, which is audibly illustrated by distinctive and exaggerated slurpy kissing sounds. This take on caricaturing RTL's and Europe n°1's directors (who are not named but clearly associated with a specific station nevertheless) in such a manner is quite revealing of the freedom of expression and creativity given to Yanne and Blanche during this show. Moreover, it displays the direction of radio stations, people rarely present on air, while placing Paris as a key location for the stations, which completes its references to geography and staff that were initiated by the sketch involving the technicians. In some ways, this joint programme can be seen as a display of the transnational dynamics and range of people involved behind the realisation of this radio production.

4 The Cool Radio Cats

One segment of this joint programme echoes the sample of audio sources studied in the first part of this chapter. This sketch stages interactions between two famous disc-jockeys of Europe n°1 and RTL, 'Hubert' and 'Rosko'. The former, Hubert Wayaffe, was, as previously mentioned, a French radio host, who started working on Europe n°1 in 1963 and was mostly known, at the time of the joint programme, for his show *Dans le vent*. The latter, Rosko, presented *Minimax*, in French, on RTL, from 1966 to 1968. Listeners would have been able to clearly identify who the characters played by Yanne and Blanche were, as the hosts call each other

[20] Translation by the author. The apparent odd choice of words is on purpose, to reflect the absurdity of the compliments said.

at the very beginning of the sketch, clearly identifying them as Hubert and Rosko. Disc-jockeys and their on-air personalities are essential in shaping a station's image; their individual performances play a part in a station's identity, while they are crucial in building and maintaining an audience.[21] By staging discussions between two famous hosts from different stations with similar on-air personalities,[22] the programme stages the relationship between the stations. Moreover, the tones used in their voices are mellow, and they speak more slowly, reinforcing this idea of intimacy, a characteristic that is even more present in night programmes.[23]

Both disc-jockeys are played as paragons of laidback 'night-owls'[24] who have high opinions of each other; they give compliments to one another over their appearance (*i.e.* about their sideburns and their shirt). Additionally, their relationship is portrayed as a love and hate dynamic; one moment they are whispering love words ('Tell me you love me') and kissing, and the other they are physically fighting. The way the programme conveys this is especially interesting; the sound moves from one ear to the other to give the impression that the characters are moving from one studio, or one station, to the other. Their fight, which is staged by the use of stereophony, provides a vivid impression of back and forth between the studios, through numerous footsteps and slapping sounds. It echoes Rosko's final comment, in which he states his surprise to have ended up in 'Europe' by mistake, before an amusingly appropriate record is played: 'Les Coups' (the blows in English) by Johnny Hallyday.

5 Cooperation Between RTL and Europe n°1 Beyond the Joint Programme

The negotiations that led to the show broadcast on the 8th May 1968 are staged and dramatised by Yanne and Blanche, but archival evidence point to a series of meetings between key figures of RTL and Europe n°1 between 1967 and 1969 discussing a potential cooperation between

[21] Chignell, *Key Concepts in Radio Studies*, p. 17.

[22] Anna Jehle, *Welle der Konsumgesellschaft: Radio Luxembourg in Frankreich 1045-1975* (Göttingen: Wallstein, 2017), p. 244.

[23] Marine Beccarelli, *La Nuit du bout des ondes. Introduction à l'histoire de la radio nocturne en France, 1945-2013* (Paris: INA Editions, 2014), p. 159.

[24] Rosko and Hubert are indeed discussing the fact that they need to have their 'goûter' (a French word for after school snack) at night as they always wake up late.

the stations. It appears that a first meeting—*réunion de travail*—between managers of RTL and Europe n°1 occurred sometime between mid-November and mid-December 1967[25] and was discussed among the directors of RTL in one of their meetings later that year.[26] A second document, from Spring 1969, summarises the ongoing discussions with Europe n°1, goes into more detail regarding concrete aspects of this collaboration,[27] and establishes the importance of following the principles mentioned in Winter 1967 as a basis. The document briefly mentions that other talks occurred in the previous spring (1968), which points to the regularity of such meetings between the two stations during the period 1967–1969, although details are scarce.

It appears that the rapprochement between the two stations was triggered by the possible introduction of publicity on the ORTF, something that would have upset the balance of France's radio landscape. If a first meeting with Europe n°1 was organised by Mr. Closon, then, by December 1967, this initiative was fully approved by the executive team, which appointed a special commission whose tasks were to pursue the first contacts with the other station, and to study all potential problems in reaching an agreement that benefitted both stations, while making sure to respect free competition. The spirit of these early contacts was the defence of private radio stations ('*la défense des stations à caractère privé*'), a phrasing that is particularly revealing of a perceived closeness, in nature, between RTL and Europe n°1. This was in opposition to the public broadcaster, which is perceived here as a threat.

The second document, from March 1969, summarises the progress made by the commission and highlights two areas of shared interest and concerns. On one hand, both stations are interested in finding solutions to reduce expenses, while on the other hand, they wished to 'form a

[25] The exact date is not mentioned in the Minutes of the 'Comité de Direction' of the 18th December 1967. As this meeting is not mentioned in the previous minutes, from 13th November 1967, it is safe to assume the meeting happened sometime between these two dates.

[26] Compagnie Luxembourgeoise de Télédiffusion, Comité de Direction, Procès-Verbaux, Séance du 18 décembre 1967, Archives RTL Group, Luxembourg.

[27] La Coopération entre R.T.L. et Europe n°1, 20th March 1969, Archives RTL Group, Luxembourg.

common front and oppose the development of State radio',[28] using again a vocabulary of conflict against a shared threat. More precisely, agreements were made in several aspects. First, regarding staff, directors of Europe n°1 and RTL agreed to exclude overbidding and headhunting, and, administrative services even exchanged information regarding salaries and compensations, with the involvement of a press agencies union, presumably to harmonise and increase the attraction of working conditions of both stations. Second, staff members of each side were regularly meeting to discuss advertisement contracts[29] in order to avoid companies wishing to advertise on commercial radio to 'exploit the competition between RTL and Europe n°1'. It is in regard to production—understood in the source as radio programmes and their promotion—that agreements were not reached. Interestingly, stations did not appear as willing to find such agreements at the time, due to, as the document states, the importance of independence in programme-making as a core activity for each station. RTL and Europe n°1 were well aware of the crucial relation between revenue and observed listenership, based on polls, making this independence in programmes on a non-negotiable factor on both ends. However, the source states, however, that the stations made a series of gestures of goodwill. For instance, RTL had regularly invited artists 'known for their connections to Europe n°1' on air, and on the 8th May 1968, both stations presented a stereophonic common show, which can be understood as proof that the rapprochement was heading in the right direction, and thus was made public.

These sources highlight the existence of concrete attempts made by both stations to reach agreements and collaborate, beyond the amusing sketches of the joint programme, though such instances were highly specific. It appears that RTL and Europe n°1 were keen to ease the competition regarding the stations' operations, while they made a clear decision to maintain their independence in terms of programmes and identity. Things became even more interesting when, despite this maintained independence, the stations felt the need to display, on air, a 'common front'. This echoes, without a doubt, the opening remarks of Jean Yanne and Francis Blanche who talked about 'radio war and peace'.

[28] La Coopération entre R.T.L. et Europe n°1, 20th March 1969, Archives RTL Group, Luxembourg.

[29] Jehle, *Welle der Konsumgesellschaft*, p. 354.

Perhaps part of the reason to convey a 'common front' through the joint programme was to be noticed by listeners, as well as by the 'shared enemy', otherwise known as the French State.

The joint programme is therefore unusual, unique, and is an oddity, and belongs to a longer history of collaboration and 'rapprochement' between the two commercial stations. Moreover, this joint programme is fascinating in regards to the question of intimacy, a rather well-known aspect of radio studies[30] that is often present on radio programmes. It comes into question here because of the show's unusual characteristics. Intimacy, which may be understood as a sense of personal closeness between listeners and the stations,[31] is particularly prevalent in radio drama and with disc-jockeys and presenters. This entire show dramatises the relationship (competition and collaboration) between RTL and Europe n°1, as if the stations were incarnated into a series of characters, inviting listeners to use their imagination.[32] Moreover, most characters present throughout the pieces are famous hosts and disc-jockeys, meaning that they are well-known to the audience and function as key characters that help to build this intimacy. This intimacy is nuanced, however, due to the lack of personal address to the audience, which is another factor of intimacy building. Apart from the advertisements and the pre-programme section, which is entirely dedicated to listeners, the rest of the joint programme almost never addresses its audience, for most of the dynamics present within the dialogue are going back and forth between two characters or the two hosts. Countering this, the beginning of the programme insists on the historic aspect of this joint programme and characterises it as a media event, in which the listeners are invited to participate. In many ways, this may have given an impression of being close to these stations, of feeling included in the entire event, even if the direct addresses are missing.

[30] Andrew Crisell, *Understanding Radio*, 2nd ed. (London: Routledge, 1994), p. 11.

[31] Chignell, *Key Concepts in Radio Studies*, p. 85.

[32] *idem.*

References

Anderson, Benedict. *Imagined Communities: Reflections on the Origin and Spread of Nationalism*, 3rd ed. London: Verso, 2006.

Beccarelli, Marine. *La Nuit du bout des ondes*. Paris: INA Editions, 2014.

Chignell, Hugh. *Key Concepts in Radio Studies*. London: Sage, 2009.

Hilmes, Michele. "Radio and the imagined community". In *The Sound Studies Reader*, edited by Jonathan Sterne, 351–362. Abingdon-on-Thames: Routledge, 2012.

Jehle, Anna. *Welle der Konsumgesellschaft: Radio Luxembourg in Frankreich, 1945–1975*. Göttingen: Wallstein Verlag, 2018.

Legay, Richard. "RTL & Europe n°1 as central actors. The importance of mobility for commercial radio stations during the Parisian events of Mai 68". *Rundfunk & Geschichte* 3:4 (2018): 41–50.

Legay, Richard. "The Role of Commercial Radio Stations in the Media Vacuum of Mai 68 in Paris". *VIEW Journal of European Television History & Culture* 6:12 (2017): 1–11.

Maréchal, Denis. *RTL. Histoire d'une radio populaire: De Radio Luxembourg à RTL.fr*. Paris: Nouveau Monde Éditions, 2010.

Nicholson, Heather Norris. "Living on Location: Amateur Creativity and Negotiating a Sense of Place in Yorkshire". In *Regional Aesthetics. Mapping UK Media Culture*, edited by Ieuan Franklin, Hugh Chignell and Kristin Skoog, 17–34. Basingstoke: Palgrave Macmillan, 2015.

Rikitianskaia, Maria. "A Transnational Approach to Radio Amateurism in the 1910s". In *Transnationalizing Radio Research: New Approaches to an Old Medium*, edited by Golo Föllmer and Alexander Badenoch, 133–140. Bielefeld: Transcript.

PART III

Intermediality

Studying radio stations, notably in the Sixties, reveals rather quickly that many produced a 'plurimedia offer'[1] for their consumers. Europe n°1 and the English service of Radio Luxembourg were no exception and the stations worked closely with magazines, respectively with *Salut les Copains* and *Fabulous 208*. Television, to some extent, was included in this 'plurimedia offer' of commercial radio stations. Europe n°1 was involved in the short-lived experiment of Telesaar in the 1950s, the first private German television station. The CLT, the company behind Radio Luxembourg, was also responsible for Télé-Luxembourg, a commercial television station that was first broadcast in 1955. However, the focus will be in the following chapters on magazines. Photos of artists, radio hosts, and disc-jockeys, whose voices and performances were regularly heard on air, were common in these periodicals, subsequently providing visual material to the on-air programmes, meaning that magazines represent a rich source for media and cultural historians.[2] Moreover, a key characteristic of radio is its secondariness, meaning 'that the listener can easily perform some other activity (work, drive and so on) while listening and paying attention to the radio'.[3] This feature of the medium supports the idea that both

[1] Claire Blandin, "Radio et magazine: une offre plurimédia pour les jeunes des Sixties", *Le Temps des Médias* (2:21, 2013), pp. 134–142.

[2] Claire Blandin (ed.), *Manuel d'analyse de la presse magazine* (Malakoff: Armand Colin, 2018), p. 11.

[3] Hugh Chignell, *Key Concepts in Radio Studies* (London: Sage, 2009), p. 70.

radio programmes and magazines could be consumed together, feeding into one another. Magazines would provide, for instance, pictures and lyrics of artists whose records were played on commercial radio, joining sound to text and image. Therefore, a cross-media perspective that does not look at radio through a single-medium lens is necessary to fully understand this intermedial radio culture.[4] The work of the *Entangled Media Histories* network is particularly useful here to study the intermediality of commercial radio culture. The key concept developed by the network is that entanglements are 'intended or unintended, obvious or hidden, structured or chaotic interrelation(s) in space, knowledge or time',[5] which cast light on blind spots in media history. This idea of blind spots is particularly strong as the following are at the crossroads of two media worlds: magazines and radio.

Part of the complexity of studying magazines resides, according to Maria R. Prior-Miller, in defining them: 'A second challenge to studying magazines is the form. No systematic studies appear to have been done of definitions over time, but source definitions reviewed for this study demonstrate the difficulty of framing a succinct, unique and ahistorical description of its essence'.[6] Historians and social scientists agree, however, on the importance, for magazines, of images and visual elements, such as photos, graphs, and other illustrations,[7] which provide fascinating material to the study undertaken here. If images are a key feature of magazines, it is always in relation to the text, and both systematically go hand in hand and are inseparable.[8] One of the challenges encountered by scholars of

[4] Richard Legay, 'Studying 1960s commercial radio', in K. McDonald & H. Chignell (eds.), *Bloomsbury Handbook of Radio* (London: Bloomsbury, 2023), pp. 367–382.

[5] Marie Cronqvist & Christoph Hilgert, "Entangled Media Histories. The Value of Transnational and Transmedial Approaches in Media Historiography", *Media History* (23:1, 2017), pp. 130–141.

[6] Marcia R. Prior-Miller, "Magazine Typology. Using Operational Classification Technology", in David Abrahamson & Marcia R. Prior-Miller (eds.), *The Routledge Handbook of Magazine Research. The Future of the Magazine Form* (Abingdon: Routledge, 2015), pp. 22–50.

[7] Jean-Marie Charon, *La Presse Magazine* (Paris: La Découverte, 2008), p 4.

[8] Dominique Maingueneau, *Analyser les textes de communication*, 2nd ed. (Paris: Nathan, 2000), p. 119.

periodical studies is the need to rely on interdisciplinarity,[9] as magazines are themselves at the crossroads of multiple fields.

In addition to entanglements, two notions are needed to analyse a corpus of magazines: intertextuality and intermediality. Despite being interlocked with one another, they have different focus and therefore are used to conduct analysis on various scales. **Intertextuality** suggests seeing a text, not as a closed and autonomous object, but rather as 'a dynamic site in which relational processes and practices are the focus of analysis instead of static structures and products'.[10] In other words, through intertextuality, the researcher finds connections in a text with other preceding texts. Here, intertextuality is an ideal concept to analyse magazines by looking into editorials, letters, and articles, and to see how specific sentences and references create bridges with radio. Secondary and tertiary texts can be added to explain how intertextuality interacts with a primary text.[11] It concerns 'references *within* media' and 'conversation *about* media'. This is of particular interest here, as it applies well to *Fabulous 208* and *Salut les Copains*, which, among other things, informed readers on commercial radio stations. **Intermediality** looks into the relations and interactions between media. It could be defined as a 'concept that points at processes of transfer and migration, between media, of format and content',[12] and developed to complement intertextuality[13] by looking at larger interactions between media. The meaning of 'text' evolved to encompass 'any aesthetic production which forms a "fabric" of words, sounds, pigments or images'.[14] 'Text' is indeed understood in a large sense, as 'humanistic scholarship gradually came to recognize

[9] M. Van Remoortel, K. Ewins, M. Koffeman & M. Philpotts, "Joining Forces: European Periodical Studies as a New Research Field", *Journal of European Periodical Studies* (1:1, 2016), pp. 1–3.

[10] Maria Jesus Martinez Alfaro, "Intertextuality: Origins and Developments of the Concept", *Atlantis* (18:1–2, 1996), pp. 268–285.

[11] John Fiske, *Television culture* (London, UK: Methuen, 1987).

[12] André Gaudreault, *Du littéraire au filmique: Système du récit*, (Malakoff: Armand Colin, 1999), p. 175.

[13] Jürgen E. Müller, "L'intermédialité, une nouvelle approche interdisciplinaire: perspectives théoriques et pratiques à l'exemple de la vision de la télévision", *Cinémas* (10:2–3, 2000), pp. 105–134.

[14] Éric Méchoulan, "Intermédialités: le temps des illusions perdues", *Intermédialités/Intermediality* (1, 2003), pp. 9–27.

texts as any vehicle of meaning, for example images, conversations, or everyday artifacts.'[15] A useful definition of intermediality used subcategories to develop the concept[16]: (1) as 'medial transposition', such as adaptations from one medium to another, (2) as 'media combination', such as the combination of different media, (3) as 'intermedial references', when a medium 'thematizes, invokes, or imitates elements or structures of another'.[17] Therefore, intermediality applies to magazines as well as programmes broadcast by commercial radio stations.

[15] Klaus Bruhn Jensen, "Intermediality", in K. B. Jensen & *al.* (eds.), *The International Encyclopedia of Communication Theory and Philosophy* (Chichester: John Wiley & Sons, 2016), pp. 972–982.

[16] Irina Rajewski, *Intermediales Erzählen in der italienischen Literatur der Postmoderne von den giovani scrittori der 80er zum pulp der 90er Jahre* (Tübingen: Gunter Narr, 2001).

[17] Irina Rajewski, "Intermediality, Intertextuality, and Remediation: A Literary Perspective on Intermediality", *Intermédialités/Intermediality* (6, 2005), pp. 43–64.

CHAPTER 8

Fabulous 208, Salut les Copains, and Their Links to Commercial Radio

Radio and press have a long interconnected history, and there are a few publications that have promoted radio listings in their pages.[1] However, *Fabulous 208* and *Salut les Copains* are of particular interest. The closeness of these two magazines with commercial radio stations, and—in the case of *Salut les Copains*—with a specific show, whose host oversaw a large variety of publishing activities, makes them especially relevant to the study undertaken in this book. Although they shared some similarities, the two magazines were not entangled with commercial stations in the same way. Relying on these two magazines as counterparts is supported by Chris Tinker's approach, who built a comparative analysis between France and Britain's youth history. Both magazines were important in the shaping of a celebrity culture, and were inspired by 'star-oriented youth magazines' from the US,[2] furthering the connections between them.

[1] For instance, *Radio Times* in the British context, *Mon Programme* or *La Semaine radiophonique* in France, which were among the best selling weekly magazines in the late 1930s. See C. Delporte, C. Blandin & F. Robinet, *Histoire de la presse en France: XXe-XXIe siècles* (Malakoff: Armand Colin, 2018), p. 110. In Germany, *Hörzu* was also a popular example. See Lu Seegers, *Hör zu!: Eduard Rhein und die Rundfunkprogrammzeitschriften (1931–1965)*, 2nd ed. (Potsdam: Verlag für Berlin-Brandenburg, 2003).

[2] Chris Tinker, 'Mixed Masculinities in 1960s British and French Youth Magazines', *The Journal of Popular Culture* (7:1, 2014), pp. 84–108.

R. Legay, *The Cultural Histories of Radio Luxembourg and Europe n°1*,
https://doi.org/10.1007/978-3-031-46250-4_8

Moreover, these youth magazines have been used as sources to study their impact on society and on young readers, notably the construction of gender identities,[3] as well as the representations of black celebrities in *Salut les Copains*.[4] If the magazines are not absent from historical literature, their connections with commercial radio stations are often overlooked. Other similar magazines, either found in youth or musical press, have also been at the centre of attention of historians.[5] Furthermore, the interrelations between radio and the press exist since the advent of the former, as mentioned by Stamm in the American context[6]: 'Cementing the partnership between newspapers and radio was not a seamless process, but there was a clear trend towards institutional and cultural harmony that took place as Americans became radio listeners during the 1920s and 1930s'.[7] A partnership which characterised 'much of media history before

[3] Tinker has shown, for instance, how youth magazines constructed masculinities in France and Britain, usually according to dominant values at the time, for instance in regards to sexual orientation. Although, it is important to note the differences by countries, notably regarding male stereotypes. In France, for instance, the image of young patriotic conscripts and the *jeune cadre dynamique* (young and active executive) dominates in the youth press. Tinker, 'Mixed Masculinities', pp. 84–108.

[4] Black artists, such as James Brown, Fats Domino and Henri Salvador were present in the magazine, however, always accompanied by strong racial stereotypes, which were also found in the photo-reportages of white artists travelling to Africa. Overall, this reinforces the visions of the community of *copains* as predominantly white. See Mat Pires, 'Les stars noires et *Salut les Copains*, 1962–1968', *Communication et langages* (n°111, 1997), pp. 59–71.

[5] Among them, Mat Pires worked, for instance, on various titles such as *Mademoiselle Âge Tendre*, *Rock & Folk* and *Nous les garçons et les filles*, while Aline Maldener compared youth magazines in France, Germany and Britain in the 1960s and 1970s, showing a rather wide range of scholars looking into such publications as historical material. See Mat Pires, 'The Popular music press', in Dauncey & Cannon, *Popular Music in France*, pp. 77–96, and Aline Maldener, 'Fabulous consumerism? Mediale Repräsentationen jugendlicher Konsumkultur in westdeutschen, britischen und französischen Jugendzeitschriften der 1960er und 1970er Jahre', in Dietmar Hüser (ed.), *Populärkultur transnational: Lesen, Hören, Sehen, Leben im Europa der langen 1960er Jahre* (Bielefeld: Transcript, 2017), pp. 199–224.

[6] Although, Britain faced a similar dynamic. See Paul Rixon, 'The Role of British newspapers in framing the public perception and experience of European radio, 1930-1939', *Media History* (20:2, 2020), pp. 153–166.

[7] Michael Stamm, 'The Sound of Print: Newspapers and the Public Promotion of Early Radio Broadcasting in the United States', in D. Suisman & S. Strasser (eds.), *Sound in the Age of Mechanical Reproduction* (Philadelphia: University of Pennsylvania Press, 2010), pp. 221–241.

television [and] continued into the 1940s and 1950s'.[8] Such a partnership between radio and magazines also meant that they would be defined by similar characteristics.

1 History of the Magazines

Salut les Copains. The monthly magazine *Salut les Copains* was launched in the summer of 1962 by Frank Ténot and Daniel Filipacchi, hosts of the eponymous radio show on Europe n°1. Capitalising on the programme's success,[9] the magazine quickly reached high sales[10] throughout the Long Sixties—despite a controversial second issue[11]—and publication was terminated in 2006, following an ultimate drop in readership. Its sales were already decreasing after the end of the radio show in 1969, which led to the necessity to relaunch the magazine as *Salut* in 1974, furthering

[8] Michael Stamm, *Sound Business: Newspapers, Radio, and the Politics of New Media*, 2nd ed. (Philadelphia: University of Pennsylvania Press, 2016), p. 186. An argument drawing from several other works, such as Alan Trachtenberg, *The Incorporation of America: Culture and Society in the Gilded Age* (New York: Hill and Wang, 1982); Douglas B. Craig, *Fireside Politics: Radio and Political Culture in the United States, 1920–1940* (Baltimore: John Hopkins University Press, 2006); Robert Horwitz, *The Irony of Regulatory Reform: the Deregulations of American Telecommunications* (Oxford: Oxford University Press, 1989).

[9] Mat Pires even called it a 'cultural landmark'. See Mat Pires, 'The Popular music press', in H. Dauncey & S. Cannon (eds.), *Popular Music in France from Chanson to Techno* (Aldershot: Ashgate, 2003), pp. 77–96.

[10] The first issue of the magazine sold 183,927 copies, a number that increasingly grew to reach a million per issue after a year, for the July 1963 issue. See Christophe Quillien, *Nos Années 'Salut les Copains': 1959–1976* (Paris: Flammarion, 2009), p. 66. A proof of the magazine's success can be found in the fact that it sold twice as many copies per issue as two main competitors, *Mademoiselle Âge Tendre* and *Le Journal de Mickey*. See Alain Fourment, *Histoire de la presse et des journaux d'enfants (1768-1988)* (Paris: Eole, 1987), p. 375.

[11] Although the magazine narrowly escaped cancellation after its second issue, which portrayed, on the cover, a picture of Elvis Presley, from a movie, in which he was holding a knife. This cover was deemed inappropriate by a French commission working on youth press. The magazine was allowed to pursue its activities as the editor explained the picture was taken from a movie which was not restricted to people over 18 years old. In 1967, the magazine was under fire from religious authorities for a cover which portrayed the French singer Sheila in a tennis outfit, revealing her knees. Quillien, *Nos Années 'Salut les Copains'*, p. 51.

the distance from the radio station.[12] The magazine focused on popular music and celebrity culture, pushing forward artists, mostly from the French musical scene, and sometimes from Britain and the United States. In the words of Filipacchi, the point of the magazine was to 'show people what they heard',[13] therefore making the magazine an extension of the radio show. Hubert Wayaffe, one of Europe n°1's hosts, who worked on the radio show and, at times, with the magazine, talked about a 'perfect osmosis'[14] between the two media. He explained it by, among other factors, the presence of photographers working for the magazine in the studios, who covered the work of the hosts and their guests before and during the programmes. While the testimonies regarding Filipacchi and *Salut les Copains* have to be taken with caution, due to their praising nature, it is nevertheless clear that the ties between the show and the magazine were strong. They were two sides of the same coin. According to Tinker, the magazine was innovative as it mixed a number of features that reflected contemporary youth culture by the inclusion of intermedial news about records and concerts, anecdotes and biographical information on artists—whose personality and image were central, at the expense of their actual music[15]—fashion advice, current affairs and a cartoon mascot: Chouchou.[16] The magazine also included its own hit-parade, as well as lyrics of contemporary songs, and, at times, translations into French, making the magazine a platform for appropriation of British and American songs by French youth.[17]

Part of what makes this magazine different was how its editors put a great deal of effort on high-quality pictures and pullout posters, which were particularly popular. To do so, the magazine actively worked with famous photographers of the period, Jean-Marie Périer likely being the

[12] Chris Tinker, 'Shaping 1960s youth in Britain and France: Fabulous and *Salut les Copains*', *International journal of Cultural Studies* (14:6, 2011), pp. 641–657.

[13] According to Daniel Filipacchi, editor of the magazine and host of the radio show, in a radio interview. See Franck Ferrand, 'Au coeur de l'émission Salut les Copains !', *Au Coeur de l'Histoire*, Europe 1, 14th September 2012, with Daniel Filipacchi, Hubert Wayaffe & Eddy Mitchell.

[14] *idem.*

[15] Pires, 'The popular music press', in Dauncey & Cannon, *Popular Music in France*, p. 78.

[16] Tinker, 'Shaping 1960s youth in Britain and France', pp. 641–657.

[17] Blandin, 'Radio et magazine: une offre plurimédia', pp. 134–142.

most prominent of all. Born in 1940, Périer started his career as an assistant for Filipacchi, then a photographer for magazines such as *Marie-Claire*,[18] *Jazz Magazine*[19] and *Paris Match*.[20] After his return from his military service in Algeria, Périer joined Filipacchi as a photographer for the newly launched *Salut les Copains*.[21] The pictures he took during his collaboration with the magazine are among the most iconic of the Sixties. They played a crucial role in shaping the image of popular culture, mostly in the French-speaking world. Périer, whose work is still celebrated,[22] photographed a large number of celebrities, including Claude François, Johnny Hallyday, Sylvie Vartan, but also Mick Jagger and Salvador Dali.

One picture by Périer in particular held significant importance for the magazine. Nicknamed '*La Photo du siècle*' (The Picture of the Century), this photograph was a group shot of 46 artists[23] from—or close to—the *yéyé* movement,[24] and a plush toy of Chouchou, the mascot of the

[18] *Marie Claire* is a monthly health, fashion and beauty magazine, published since 1937 in France, and since 1941 in the United Kingdom. The magazine aims at a female readership, and was edited from its first edition to 1974 by Jean Prouvost, who became Chairman of Radio Luxembourg in 1966 and undertook many changes in the station.

[19] *Jazz Magazine* is a music magazine dedicated to jazz, published since 1954. It was founded by Eddie Barclay, a key figure of the music world in France in the 1950s and 1960s. Frank Ténot and Daniel Filipacchi became directors of the magazine in 1956, and later launched the radio show and eponymous magazine *Salut les Copains!*.

[20] *Paris Match* is a weekly news magazine, published since 1949, and known for its focus on photography. It was also funded by Jean Prouvost, and Daniel Filipacchi worked there as a photographer at the beginning of his career. In 1974, Filipacchi bought the magazine, which was facing financial difficulties.

[21] Jean-Marie Périer, *Oncle Dan. Souvenirs* (Paris: XO Éditions, 2008), p. 177.

[22] A photographic exhibition on Jean-Marie Périer's work was held in 2019 in Paris, on the *Grande arche de la Défense*. Around 300 pictures were displayed, with a third of original works, showing the long-lasting impact of the photographer on popular culture in the 1960s and its memory until now. Adeline Fleury, 'Exposition: les sixties exquises de Jean-Marie Périer en images', *Le Parisien*, 17th January 2019, [http://www.leparisien.fr/culture-loisirs/exposition-les-sixties-exquises-de-jean-marie-perier-en-images-17-01-2019-7989615.php] (last consulted 07/03/2023).

[23] Among them, Sylvie Vartan, France Gall, Eddy Mitchell, Chantal Goya, Dick Rivers and Claude François, to quote a few names. Johnny Hallyday was also present, standing on a ladder on the last row, therefore in a more prominent position over the rest of the group.

[24] The *yéyé* movement, whose name originated from the 'yeah! yeah!' popularised by beat music, was a genre of pop music popular in France and other European countries in the 1960s. It was described by David Looseley as the 'anodyne French version of

magazine. The picture, resembling a school photography, was published in June 1966 as a double-page pullout poster in the 47th issue of the magazine, to celebrate its 4th publication anniversary. The interest of *La Photo du siècle* lies in the possibility to interpret it as a display of power[25] for the magazine, and, by extension, for the show. Périer and the magazine managed to summon a great number of popular artists to be present at the same time for the shot, therefore asserting the central position of *Salut les Copains* in the contemporary music world and youth culture. Through this picture, intermedial references can be found. The artists portrayed are known for their music and their records, which are regularly played in the radio show, while their personal lives and public images are central to the magazines' pages. With the group picture, they are assembled in a single space, capturing a form of essence of popular French music, and, as the picture was published as a pullout poster, it had the potential to be displayed in many teenagers' rooms.

Fabulous 208. *Fabulous* was launched in 1964 by Fleetway, a publishing company mostly working with comics magazines at the time.[26] The first issue, on January 18th, portrayed The Beatles on its cover, and their nickname of 'The Fabulous Four' inspired the title of the magazine.[27] The band members were portrayed, one way or another, into every issue of the magazine for the next two years.[28] Despite the focus on this specific band, the magazine was actually about popular music in general, and aimed at a teenage readership. It offered a content not dissimilar to *Salut les Copains* in the French-speaking world, notably via the pullout posters and high-quality pictures of celebrities.[29] An analysis

pop', and Mat Pires pointed out that it was the 'particular musical fascination' of *Salut les Copains*. See David Looseley, 'In from the margins: *chansons*, pop and cultural legitimacy', in Dauncey & Cannon, *Popular Music in France*, pp. 27–40; and Pires, 'The Popular music press', in Dauncey & Cannon, *Popular Music in France*, pp. 77–96.

[25] Gathering so many artists was described, by Périer, as a partially challenging feat. Quillien, *Nos Années 'Salut les Copains'*, p. 104.

[26] Tinker, 'Shaping 1960s youth in Britain and France', pp. 641–657.

[27] Chris Tinker, 'Mixed Masculinities in 1960s British and French Youth Magazines', *The Journal of Popular Culture* (7:1, 2014), pp. 84–108.

[28] *Fabulous* was slightly pricier than other similar magazines, although it offered pictures of better quality. See Jon Savage, 'The Magazine explosion', *The Guardian*, 6th September 2009, [https://www.theguardian.com/music/2009/sep/06/sixties-60s-pop-magazines-beatles] (last consulted 20/05/2020).

[29] Tinker, 'Shaping 1960s youth in Britain and France', pp. 641–657.

of the magazine from a graphic point of view revealed how it brought together readers and celebrities in a sense of closeness.[30] The magazine joined forces with the English service of Radio Luxembourg in 1966, and was therefore renamed *Fabulous 208*, in reference to the station's wavelength.[31] In the first joint issue, in June 1966, the cover of the magazine clearly advertised this union, stating, in the middle of the page: 'Great news! Radio Luxembourg and Fabulous sensational link-up!', and, under it, in smaller print: 'NOW Official Programme Weekly for Luxembourg 208'.[32] Among the series of pictures printed on the cover—mostly artists either performing a song or being interviewed in front of a microphone—two are of particular interest here. On the left, a picture of a radio antenna, and, on the right, a portrait of Radio Luxembourg's disc-jockey Jimmy Savile[33] in a studio, both clearly inscribing the presence of the station in the magazine. The magazine reached sales of around 300,000 copies per issue, a high number even though significantly lower than its French counterpart *Salut les Copains*.[34] The sales dropped, however, in the 1970s and publication stopped in 1980. The content of *Fabulous 208* and *Salut les Copains* resembled one another in many ways, however, the English-language magazine published a variety of serials, a feature not found in the French counterpart. A serial, in this context, refers to the process of publishing a literary work in a series of issues of a periodical,[35] such as a magazine like *Fabulous 208*. There were, on average, two or three serials, spread over one or two pages, in each issue of *Fabulous 208*. Among them, titles like 'The World of Mandy Bennett. A week by week diary', 'The Painter and the Private Eye', 'One of Those Things'

[30] Paul Jobling & Richard Crowley, *Graphic Design: Reproduction and Representation since 1800* (Manchester: Manchester University Press, 2003), pp. 218–219.

[31] This rebranding already points at an interesting entanglement between the two media, as the new title includes directly a reference to the technical aspect of the commercial station. 208 was indeed a common way to refer to the station.

[32] *Fabulous 208*, 4th June 1966.

[33] Jimmy Savile (1926–2011) was a British disc-jockey and television personality. He started his career in 1958 at Radio Luxembourg, and later joined the BBC where he presented the famous *Top of the Pops* on television. He worked on BBC Radio 1 from 1968. Following investigations after his death, the police concluded he was a predatory sex offender, with hundreds of victims.

[34] Tinker, 'Shaping 1960s youth in Britain and France', pp. 641–657.

[35] Laurel Brake & Marysa Demoor (eds.), *Dictionary of Nineteenth Century Journalism* (Ghent: Academia Press, 2009), p. 567.

and 'The Girl from Lady Lane', could be read, often focusing on a main female character. Although the popularity of serials was not as strong in the 1960s as it was before, relying on such a format had always been a way to support the commercial venture of a magazine.[36] It also anchored *Fabulous 208* in the long history of British periodicals, inscribing it in a rather traditional format.

2 Shared Features Between Programmes and Magazines

Commercial radio stations developed strong links with youth magazines *Salut les Copains* and *Fabulous 208*. Particularly relevant throughout the 1960s, this entangled history between the two media operated on multiple scales. It is now necessary to look at them in a more detailed and precise way. Through a micro-analysis of specific elements (*e.g.* articles, sections), this chapter sheds light on these connections by unearthing tangible intertextual and intermedial elements found throughout issues of the magazines. Based on these elements, five categories were picked to classify the interrelations between magazines and commercial radio, as they are features found in both media. These features—playfulness, commercialism, localism, transnationalism, and interactivity—are a direct reference to categories created in the previous chapters on the sound of commercial radio stations in the 1960s. As such, they offer a clear framework to highlight the connections between the on-air programmes and the pages of the magazines studied in this chapter. Before moving into an in-depth look into these categories, it is important to stop on the most striking presence of on-air broadcast in a magazine: its radio programme listings. Indeed, when *Fabulous* teamed up with the English service of Radio Luxembourg, this new link between the two media was advertised, as mentioned in the previous chapter, by a cover page celebrating it. From then on, the magazine included in the first pages of each

[36] 'Serialization was part of the basic grammar of serials, as structure designed to assure continuity of sales by linking one issue, thus one sale, to another; in this way the "present" refers back to the past number as well as anticipating the future number', Laurel Brake, 'Aestheticism and Decadence: *The Yellow Book* (1894–7), *The Chameleon* (1894), and *the Savoy* (1896)', in Andrew Thacker & Peter Brooker (eds.), *The Oxford Critical and Cultural History of Modernist Magazines. Volume I: Britain and Ireland, 1880-1955* (Oxford: Oxford University Press, 2013), pp. 76–100.

issue, the weekly programme of the shows broadcast on 208. By doing so, the magazine ostensibly displayed[37] its connections with the commercial radio station and took on the role of *presse de programmes*.[38] In the British case, *Radio Times*, published in-house by the BBC,[39] was the main magazine which provided radio and television listings, however, it excluded the commercial station Radio Luxembourg. It was then through *Fabulous 208*, another commercial venture, that the station had its programmes displayed in the press therefore fighting its marginal status.

By looking at the English service of Radio Luxembourg and its links to *Fabulous 208*, the programmes published weekly offer another interesting interrelation. The importance of the temporal rhythm of radio to construct bridges between radio's hourly, daily, and weekly cycles and everyday lives[40] has to be kept in mind here. With *Fabulous 208*, this temporal rhythm is reinforced by being printed in the pages of the magazine. The listings of programmes indeed further this structural potential of media, as mentioned by Gripsrud: 'broadcast media form a "normal" time scheme for "normal" lives, they structure each day and give each day of the week a special character'.[41] Therefore, it appears that listings of programmes present in *Fabulous 208* after the magazine joined up with the commercial station were one of its key features, and offered direct insights into the entanglements operating between the two media. Firstly, by displaying a promotional platform for the station's programmes, and secondly, by furthering the structural power of radio on hourly, daily, and weekly cycles. These listings worked on both rhythms: the one of radio, as they displayed, hour by hour, Radio Luxembourg's disc-jockeys' programmes, as well as the rhythm of magazines. Published weekly, *Fabulous 208* only gave prospective programmes for seven upcoming days,

[37] The listings of programmes are filled with intertextual references to the radio station, including its wavelength (208 metres) and names of the programmes and the disc-jockeys.

[38] To use the French term designating the large body of periodicals offering listings and other content related to the audiovisual media. Blandin, *Manuel d'analyse de la presse magazine*, p. 309.

[39] Olivia Thomson, 'The History of Radio Times', *Radio Times*, 17th July 2013. [https://www.radiotimes.com/news/2013-07-17/the-history-of-radio-times/] (last accessed 30/06/2020).

[40] David Hendy, *Radio in the Global Age* (London: Polity Press, 2000), p. 178.

[41] Jostein Gripsrud, *Understanding Media Culture* (London: Arnold, 2002), p. 24.

inferring the 'elasticity' of the magazine's relation to time.[42] Programme listings of Radio 208 published in *Fabulous 208* were then at the crossroads of magazines and commercial radio, a sort of media combination between the two.

Irreverence and Playfulness. As described previously, commercial radio stations in the Long Sixties offered many programmes in which irreverence and playfulness played a key role. Such a defining feature of the sound of commercial stations seems, however, not to have been fully transferred onto the magazines. More precisely, if the playfulness of 208's disc-jockeys and hosts could be found in the pages of *Fabulous 208*, the *Salut les Copains* magazine appears to have been more 'serious', in opposition to the tone of Europe n°1 at the same period. The difference in tone between the magazines can be explained by two elements. The magazine *Salut les Copains* was an extension of the eponymous radio programme, which, arguably, did not embrace characteristics of irreverence and playfulness to the same extent as other programmes on Europe n°1.[43] Therefore, the magazine did not carry the tone of the entire station. Moreover, rules in France regarding youth magazines were stricter than in Great Britain, as shown by the cover of *Salut les Copains* with Elvis Presley holding a knife which caused the magazine a lot of trouble.[44] As such, the lack of playfulness as a defining feature of the

[42] Jamil Dakhlia, 'Propriétés et fonctions de la presse magazine', in Blandin, *Manuel d'analyse de la presse magazine*, pp. 51–65.

[43] In his comparative analysis of *Fabulous* and *Salut les Copains*, Tinker notes that the French magazine takes a rather supportive stance towards the Gaullist regime. Tinker, 'Shaping 1960s youth in Britain and France', pp. 641-657. Support in favour of De Gaulle's regime is not apparent in the eponymous radio show. Considering the uneasy relationship between Europe n°1 and the French government from its early days and throughout the 1960s, it is safe to assume that the supportive stance was the preserve of the magazine, and not shared by the other medium.

[44] In 1949, the French State indeed created a special commission in charge of surveilling and controlling publications destined to children and teenagers, which included a magazine like *Salut les Copains*. The commission, which still exists nowadays, consists of several members, including representatives of various ministers, a series of magistrates, elected politicians, representatives of unions of editors, artists, teachers, and, interestingly, a mother and a father, picked by family associations. This commission is charged with making sure that youth publications are not promoting, in any form—and among other things—organised crime, lying, laziness, cowardliness, and debauchery. The prerogatives of the commission are important and the potential consequences for editors who would not abide are strict. See *Loi n°49-956 du 16 juillet 1949 sur les publications destinées à la*

Salut les Copains magazine is not solely due to a different editorial line, but also to a different national context.

On the contrary, many playful and humorous comments, pictures, and anecdotes can be found in the pages of *Fabulous 208*, and, interestingly, most of them were made at the expense of Radio Luxembourg's disc-jockeys. This is particularly interesting while researching shared features between the two media, as it means the characteristic irreverence of the English service of Radio Luxembourg transferred in *Fabulous 208* via its disc-jockeys. They were already developing this irreverent and playful image on air, integrating it into their programmes and as an aspect of their radio personalities. This shared feature of irreverence and playfulness can be traced, in the magazine, through intertextual elements found in both texts and images of *Fabulous 208*. For instance, in a 1967 issue of *Fabulous 208*, a brief article explains, with humour, that Jimmy Savile could not be heard on the station as he was sick, and afflicted with jaundice.[45] Accompanying this text, a rather large picture—taking almost a quarter of the page—portrays Savile in bed, sleeping, with two bowls of fruit on his chest and stomach, and grapes on his forehead. Two unidentified people, dressed in black, as in mourning, are sitting beside him. This reinforced the tension between the seriousness of disease and mourning with the odd presence of fruit and the overall light tone of the article. This portrayal of a famous voice of the station, in a humouristic and, frankly, slightly absurd manner, fits in well with the self-mockery disc-jockeys played with, both in print and on air.

The self-mockery and playfulness of the radio hosts, and, by extension, of Radio Luxembourg, was best incarnated,[46] at least in the pages of *Fabulous 208*, by Tony Prince, one of the resident disc-jockeys,[47] who

jeunesse, on legifrance.gouv.fr, (https://www.legifrance.gouv.fr/affichTexte.do?cidTexte=LEGITEXT000006068067&dateTexte=20100817), last consulted 07/03/2023.

[45] *Fabulous 208*, 3rd June 1967.

[46] It is necessary to remember that the person 'speaking' in a magazine is always a construction, and the actual person only represents themselves, often via general characteristics (such as playfulness in the case of Tony Prince). See François Provenzano, 'Le Discours de la presse magazine', in Blandin, *Manuel d'analyse de la presse magazine*, pp. 85–104, & Ruth Amossy, *La Présentation de soi. Ethos et identité verbale*, (Paris: Presses Universitaires de France, 2010), p. 7.

[47] Some of the hosts of the English service of Radio Luxembourg were living in the Grand Duchy and worked directly in the studios of the Villa Louvigny, in the heart of

wrote a weekly section.[48] In one of those, he told the story of how he went, with two other hosts of the station—David 'Kid' Jensen and Paul Burnett—for horse-riding lessons, which, according to him, did not go well.[49]

The story is presented as a parody of a photo-reportage over two pages, with a comment from one of the horses, and a few—rather unflattering—pictures of the radio men horse riding. For instance, in one of them, the trio can be seen clearly in pain after the session, leaning against a fence with their legs arched. This is based on a traditional representation of male celebrities on horseback, built on the manly image of the cowboy, something not unusual in the 1960s.[50] Even *Fabulous 208* itself engaged in this type of representation, such as in a 1967 issue, which contained another colour photo-reportage, over a few pages, about the British band The Young Idea riding horses.[51] Therefore, based on these intertextual references, it could be argued that Prince's photo-reportage parodied not only the trend of photographing male celebrities on horseback, but *Fabulous 208* itself, inferring a sense of irreverence towards the magazine. Moreover, on a few occasions, *Fabulous 208* published sections dealing with news from Luxembourg (both the country and the station), which parodied, or were inspired by, newspapers. These sections copied the format and general style of a newspaper, and titled themselves 'Luxy Times' or 'The 208 Mirror', creating clear intertextual links to the wider British press within the pages of the magazine.[52]

As mentioned, pictures played a crucial role in both magazines. In *Fabulous 208*, they were regularly used to portray disc-jockeys as another

the capital. They appear to have had a more prominent place in *Fabulous 208* than other disc-jockeys.

[48] This section written by Prince was often written as a letter to readers, although it could have different formats, and would usually inform them about what was happening to the resident disc-jockeys of the station. In each case, the overall tone was never serious.

[49] *Fabulous 208*, 12th July 1969.

[50] And this was not only in Britain. In France, for instance, Johnny Hallyday was regularly photographed riding a horse (i.e. *Salut les Copains*, n°11, June 1963), in an imagery inspired by the Far West. Interestingly, he also sang, in 1963, a western-inspired song—'Les Bras en croix'—in which the lyrics tell the story, in the first person, of a cowboy in Oklahoma, suggesting a permeability of this influence in both musical and visual worlds.

[51] *Fabulous 208*, 26th August 1967.

[52] *Fabulous 208*, 15th January 1972 & *Fabulous 208*, 9th June 1973.

kind of celebrity. They offered, at times, a less serious take on the lives of the 208 hosts, picturing them wreaking chaos at a bowling alley, for instance.[53] On another occasion, *Fabulous 208* published a one-page photo-story in which two of the resident disc-jockeys, Tony Prince and David 'Kid' Jensen, are trying, in vain, to play a tape sent by a listener.[54] Five photos picture them struggling and gradually getting themselves covered in more and more loose tape, while captions of the photo-story report their dialogue, in which they ultimately decide to give up. This example shows again this playfulness displayed by the radio men, as well as the variety of media used to portray it, as the photo-story is rarely used in the magazine.

The examples of Tony Prince's pieces in *Fabulous 208* are particularly rich in instances of irreverence, not only for their photographs but also for the texts he wrote. In 1970, he reported on a trip he took to Ireland, specifically on the second part of the journey back: the flight from London to Luxembourg. Prince insists on his tiredness to explain he 'must have been the only passenger who didn't realise that Luxembourg's Grand Duke and Duchess were on board in the V.I.P. compartment', to which he adds: 'I wonder if our Grand Duke (who is the equivalent to a country's King), realised that he'd travelled with a British Prince? A very, very tired and happy Prince at that!'. The 'British Prince' mentioned in this extract is obviously Tony Prince himself, who went by the nicknames of 'Royal Ruler' and 'Regal Ruler'.[55] This attitude to refer to oneself as royalty is particularly playful,[56] and could be seen as borderline irreverent when meeting actual monarchs. The image of Prince as the 'Royal Ruler' was not limited to *Fabulous 208*, as it actually originated from his radio work and on-air personality, drawing clear intermedial connections between radio and magazine, through the image of one disc-jockey and occasional writer for *Fabulous 208*. These connections between the on-air and in-print personality of Tony Prince went even further. The short report on the flight mentioned previously was concluded in the following way: Thanks Great Britain and Eire. See you again soon! Rule Britannia

[53] *Fabulous 208*, 30th May 1970.

[54] *Fabulous 208*, 10th January 1970.

[55] Likely in reference to Prince, his on air last name.

[56] In a similar fashion, Mike Pasternak, better known as Rosko, went by the nicknames of 'Emperor' or 'Président'.

Dee Dum Dee Dum Dee Luv. And a Happy New Year to you too luv luv, Tony'.[57] Written as a personal letter to readers, this example shows how playful closings could further the image of irreverence by circumventing basic rules of the English language and norms of print media. This habit of playing with language was common for Prince, who wrote, on another occasion, the following greetings: 'Hiya m'fave rave right knights, yer Royal Ruler here to lay pen and ink on you once again', and signed 'Love', but written without the letter 'O' and with no more than eighteen 'Ls'.[58] Not only do these examples show the latent irreverence and playfulness of one disc-jockey in the magazine, they directly echo the way he spoke while hosting programmes on Radio Luxembourg. Therefore, this is where the intermedial entanglement took a particularly interesting aspect, as the printed words of the magazine reflected the spoken words of the commercial station, tying together, through a form of intertextuality, the two media. This occurred through one man, who operated in both worlds, and the personality he developed, also spread between the station and the magazine. This shows the important role that radio celebrities had: to carry and transfer features across media. Through their personalities, the transfers of such features are facilitated as playfulness and irreverence are attached to their image, but can be expressed in various media.

Commercialism and Advertisement in Magazines. Radio stations like Europe n°1 and Radio Luxembourg are defined by their commercial aspect, meaning they are run as businesses, in opposition to public service.[59] Their use of advertisement on air, as a way to be funded, is therefore the most tangible expression of this commercialism. By doing so, they shared this commercial feature with magazines.[60] Indeed, according to Jamil Dakhlia, magazines are excellent advertising media, due to the importance of visual components, as well as knowing the socio-demographic characteristics of their targeted readership and being able to

[57] *Fabulous 208*, 10th January 1970. This is a direct transcription of what is written.

[58] *Fabulous 208*, 4th October 1969. Once again, this is a literal transcription.

[59] Chignell, *Key Concepts in Radio Studies*, p. 114.

[60] Focusing on France, Delporte, Blandin and Robinet have shown how important the 1960s–1980s period was for advertisement in the press, where it faced exponential development. Delporte, Blandin & Robinet, *Histoire de la presse en France*, pp. 228–239.

exploit such knowledge.[61] Interestingly, he also noted further that advertised products were often '*coulés*' (cast, or poured) into the pages of the magazines, through a close following of stylistic norms of the periodical. These practices of relying on adverts were shared by both media, offering an interesting angle to study how entangled they were, especially as both put efforts into moulding these adverts to follow the specifics of their own medium.

Moreover, many adverts in the magazines referred to the radio world, by promoting the stations, their programmes, as well as the technical objects related to commercial radio culture: transistor sets and record players notably. Adverts for these types of products were numerous in both *Salut les Copains* and *Fabulous 208*, implicating the closeness of the links operating between the magazines and the radio stations. They also played—as did magazine adverts—a part in promoting a larger commercial logic and supposed the compliance with a lifestyle and consumerist ideology.[62] In this case, it revolved around the consumption of radio programmes, magazines, records, and all the technology associated with it.

In an issue of *Salut les Copains* from 1966, an advert for Kodak cameras can be found.[63] Interestingly, it featured a portrait of the magazine's famous photographer, Jean-Marie Périer. Well-known to readers of the magazine, the fact that the brand of cameras used his image to promote their product also shows an example of media combination—via advertisement—between the magazine and photography. While some adverts were clearly identifiable as such, the commercial promotion of other products—such as records of an artist whose name comes up regularly in an issue—are more subtle; suggesting this invisible and visible tension at play between the entanglements of media histories.[64] The few

[61] Dakhlia, 'Propriétés et fonctions de la presse magazine', in Blandin, *Manuel d'analyse de la presse magazine*, pp. 51–65.

[62] Dakhlia, 'Propriétés et fonctions de la presse magazine', in Blandin, *Manuel d'analyse de la presse magazine*, pp. 51–65.

[63] *Salut les Copains*, n°46, May 1966.

[64] One of the most famous examples of an analysis of such a subtle, invisible, promotion of a lifestyle was made by Barthes, who analysed cooking recipes in the *Elle* magazine and showed how they were used to support a vision of cooking as high-end, sophisticated, and beautiful, rather than truly culinary. Roland Barthes, 'Ornamental Cookery', in *Mythologies*, revised ed. (London: Vintage, 2009, translation by Jonathan Cape), pp. 89–91.

examples developed above show how advertisement, a feature of both magazines and commercial radio stations, operated through numerous intertextual and intermedial interrelations, bridging not only the two media together but also referencing other media and technologies. As such, they also operate as a reminder of the double nature of radio, which is both an object and a medium,[65] directly referencing a specific radio programme or a new radio receiver.

Localism and Foreignness. As shown in a previous chapter, commercial radio stations put out an image balancing between localism and foreignness, making it a key feature of their identity. By looking into the magazines published in connection with the stations, it appears that such aspects were carried onto the pages of *Fabulous 208* and *Salut les Copains*. The two notions adapted, however, to the new medium and mostly relied on the complementarity of text and image that makes the strength of magazines. In this regard, one photo-reportage published in *Fabulous 208* in June 1967 is particularly insightful. As a celebration of the first anniversary of the collaboration with Radio Luxembourg, a group of staff members of the magazine went to the Grand Duchy for a few days, to meet with the local team of resident disc-jockeys, and to explore the capital city. The group was composed of reporter Christine Osbourne, photographer Fiona Adams, and British singer Engelbert Humperdinck,[66] who posed for all the pictures taken. This photo-reportage was a central piece of the issue, as shown by pictures of Humperdinck posing around Luxembourg-city over half of the cover. The editorial even advertised it in the following way: 'This is a specially happy Fab-208. We've had a super year since we linked up with the Luxembourg lot and we've had a few trips out there, too'.[67] This insistence on the anniversary is of particular interest, especially as it took the form of a trip to Luxembourg, joining, for a few days, the differently intertwined

[65] Andreas Fickers, 'Neither Good, nor bad; nor neutral': The Historical *Dispositif* of Communication Technologies', in M. Schreiber & C. Zimmermann (eds.), *Journalism and Technological Change: Historical Perspectives, Contemporary Trends* (Frankfurt: Campus, 2014), pp. 30–52.

[66] Arnold George Dorsey (born in 1936), better known as Engelbert Humperdinck, is an English singer of pop music, whose career took off in the mid-1960s.

[67] *Fabulous 208*, 3rd June 1967.

worlds; the reporter and photographer for the magazine, the disc-jockeys for the radio stations, and the singer for popular culture and the record industry.

The journey was, in some ways, used as a dramatisation of the links uniting the station and the magazine, therefore offering many insights into the ways it was portrayed to its readers and listeners. Part of the reportage included photos and comments regarding the journey, by plane, to Luxembourg and back, while the editorial reminded readers of the functioning of the stations: 'Some people still don't realise that there really is a landlocked country in the middle of Europe with four dishy Djs putting over the 208 programmes and seeing that the tapes which are recorded in Hertford Street, London, are flown out on time and are broadcast in the correct place on the schedule'. Both elements help shape the transnational image of Radio Luxembourg, and the constant links tying the station with London. The photo-reportage also portrayed Luxembourg, understood here as an umbrella term for both the station and the city. It included a picture of the inside of the studios at the Villa Louvigny, showing two of the disc-jockeys—Don Wardell and Colin Nicol—interviewing Engelbert Humperdinck on air. Meanwhile Osbourne, the reporter, described in a few words what her experience of the station at night was. Such elements published in the pages of *Fabulous 208* are of particular importance to shape, in the mind of the reader-listeners, the way radio was made. It gave insights into the work behind it, as well as the transnational dimension specific to this commercial radio station.

The second attempt to portray Luxembourg concerned, as mentioned, the capital city of the Grand Duchy. Throughout the reportage, photos and text include—rather vague—references to well-known landmarks of the city. Among them, Osbourne mentions 'the new—bright red! — bridge', which refers to the Grand Duchess Charlotte-Bridge, built in 1965, and the 'town prison', which was located in the Neumünster Abbey at the time. A photo of Engelbert shows him looking down a valley, and, in the back, another well-known sight of the city, its viaduct, overhanging the Petrusse Valley. All of these elements, in addition to anecdotes regarding the various languages spoken in the country, give the photo-reportage a strong local dimension, implementing the links to the Grand Duchy in the magazine, in a similar way to what was done on air. All these local references reinforce the foreign aspect of Luxembourg, by showing its otherness to a mainly British readership. And, as shown in a previous chapter, foreignness could be associated with a form of liberty, in direct echo to the nature of commercial broadcasters. As such, a photo-reportage like the one studied above does not only portray the transnational link bridging Luxembourg to Britain, it also suggests the supposed freedom resulting from this link.

Both the country of Luxembourg and its eponymous capital city were incorporated in the pages of *Fabulous 208*. The example of the photo-reportage developed above is a rather lengthy one, and most of the references to the Grand Duchy were shorter ones. Such references were already present in the soundscape of the English service of Radio Luxembourg and also appeared in the pages of the magazine. For instance, in an anecdote developed previously, Tony Prince shared his journey back to Luxembourg from Dublin and via London, as well as his unexpected encounter with Luxembourg's Grand Duke.[68] The Grand Duchy was also portrayed as an exotic destination, due to the presence of the station. The magazine even organised a group trip to Luxembourg for its readers. Three-days long and all-inclusive, the trip was advertised as some sort of holiday, during which participants would discover Luxembourg; the station, its studios, meet the disc-jockeys, and enjoy a night in a local discotheque.[69] To promote the trip and encourage people to participate—for a fee, obviously—the magazine advertised it over two pages, detailing the programme hour by hour. The transnational journey, crossing England, the Channel, France, and, finally, Luxembourg, is also romanticised, through pictures of things as mundane as the coach used on the continent, and a map detailing the journey. In a similar way to the ones in the radio programmes, these references shaped the transnational image of the station and, by extension, of the magazine, by reminding readers of its presence on the continent, despite a strong British anchoring. Moreover, many issues of *Fabulous 208* included a regular section on news from the United States, about the music and movie scenes. This section added a transatlantic layer to the image displayed by the magazine, as the news was said to be directly imported from Hollywood and was sometimes romanticised in the magazine.[70]

68 *Fabulous 208*, 10th January 1970.

69 *Fabulous 208*, 16th January 1971.

70 In 1970, a few lines in the magazine explaining the news from the United States were obtained at the last minute via phone, referring to the transnational communication network, implicitly adding them to the transnational image of the magazine and the

On the other hand, *Salut les Copains* also included numerous articles revolving around journeys to foreign countries, to further an image of internationalism. For instance, in the same issue from December 1962,[71] the magazine featured two photo-reportages which portrayed Cliff Richard[72] and Petula Clark.[73] The first one is photographed all over London, which is easily recognisable thanks to the presence of many landmarks of the city, as well as 'bobbies', which clearly resonate with French-speaking readers as a form of Britishness. The second one focuses on the Franco-English image of Petula Clark, which is illustrated by details of her life in Paris and in Britain, as well as photos of her cottage in England and of one of her train journeys across the Channel. As a whole, these two photo-reportages showcase this omnipresent image of *Salut les Copains* by representing the internationality of the celebrities with whom the magazine collaborated, and by romanticising the journeys to foreign countries. In his article on the representations of Black celebrities in *Salut les Copains*, Mat Pires notes that the magazine featured many photo-reportages of White celebrities travelling to various locations in Africa, almost systematically accompanied, in the texts and in the photos, by racial stereotypes.[74] It is interesting to note this aspect of the foreignness in the magazine, as it is less striking in the radio show, therefore inferring a difference between the two media.

Interactivity. *Salut les Copains*—as a dual product, both a radio show and a magazine—is of particular interest in the development of a group of consumers, both listeners and readers, of both male and female, and usually in their teenage years.[75] Addressed as the *copains*, or 'mates', they appear close to what Benedict Anderson called an *imagined community*,

station: 'Whew! We nearly had a blank space where Janey Milstead's Pop Press from Hollywood should be! But at the last minute we managed to get her on the telephone—transatlantic of course!'. See *Fabulous 208*, 10th January 1970.

71 *Salut les Copains*, n°5, December 1962.

72 Cliff Richard (born in 1940) is a British singer, mostly known for his collaboration with the backing British rock band *The Shadows*.

73 Petula Clark (born in 1932) is a British singer, whose career started on the BBC during World War II, and took off in the 1960s and 1970s. She sang records in multiple languages, and especially in French and English.

74 Pires, 'Les stars noires et *Salut les Copains*', pp. 59–71.

75 Blandin, 'Radio et magazine: une offre plurimédia', pp. 134–142.

a concept he used to describe the birth of nationalism worldwide.[76] This idea was later reused in other frameworks, notably by Michele Hilmes, who pointed out that imagined communities could also be shaped by the ritualised consumption of radio, where Anderson focused on the newspaper.[77] Therefore, a close look at the *Salut les Copains* magazine, which, by its nature, is close to a newspaper and to the radio stations, is an interesting gateway into the inclusion of listeners-readers, and the shaping of their community. The shapes of this inclusion varied but all shed light on entanglements binding the radio show with the magazine beyond the historical and financial one developed previously. This act of inclusion of the readers is of significant interest to the research undertaken in this chapter, as they gave listeners a possibility to express themselves and exchange with each other and with the team behind *Salut les Copains*. In return, their inclusion was also displayed in the magazine, allowing it to shape its image accordingly.

Salut les Copains, like other magazines and newspapers at the time, published, in every issue, letters, or extracts of letters sent by its readers. While the format evolved over time and the number of correspondences published in each issue changed, the overall dynamics stayed the same. One key feature running through all correspondence is that they were addressed to 'Daniel' (Filipacchi), the editor of the magazine, and host of the eponymous radio show, making him the central figure, the 'main *copain*' of the community, a position he already held in the radio show, suggesting a figure spread over both media. The majority of letters published in the pages of the magazine were rather positive, praising articles and photo-reportages published in previous issues. Some were also made by readers who wanted to share personal anecdotes, usually linked to a concert or a record. These letters helped shape a sense of seriality between the issues of the magazine, by mentioning past issues, and they

[76] An 'imagined community', according to Anderson, who coined the term in the 1980s, reflects on how a group of individuals think of themselves as a nation, despite the fact they 'will never know most of their fellow-members'. They still belong to a community due to the 'deep, horizontal comradeship' developed through ceremonies shared by thousands, such as the reading of newspapers. Benedict Anderson, *Imagined Communities: Reflections on the Origin and Spread of Nationalism*, 3rd ed. (London: Verso, 2006), p.6.

[77] Michele Hilmes, 'Radio and the imagined community', in Jonathan Sterne (ed.), *The Sound Studies Reader*, (Abingdon-on-Thames: Routledge, 2012), pp. 351–362.

also displayed a supposedly international readership of the magazine.[78] Quite a few letters, on the other hand, were particularly negative. Letters from a reader, who signed with the pseudonym 'Billy Gluboh', were published in *Salut les Copains* on a few occasions and were especially aggressive.[79] Interestingly, this correspondence often sparked reactions from other readers, which could be found published in the following issues,[80] reinforcing the intertextual links between the letters.According to Jean-Marie Charon, who based his idea on the concept of *contrat de lecture*[81] (reader's contract) developed by Eliseo Veron, magazines stood out among the press, because their content was based on their readers rather than on current affairs.[82] More precisely, editors of a magazine base their offer on the taste, experience, and expectation of their readers, who may or may not adhere to the content in return. Therefore, letters published in *Salut les Copains* could be seen as a display, from the magazine, of their alleged intention of taking their readers' opinion into account. Indeed, most of the letters concern comments on articles published in previous issues, as well as on movies, concerts, television, radio, and records, making the letters intermedial objects within the magazine. As such, the letters created a sense of interactivity and inclusion of the readers, in a way close to what the commercial radio station was

[78] Indeed, many letters were sent by readers abroad, such as 'Dave' from Southampton, in England, who believes French singers like Eddy Mitchell and Richard Anthony would be successful in England, and, in the same issue, Isabelle, from Madrid, in Spain, writes a letter in Spanish, and which she explains she is the secretary of the Johnny Hallyday Club of Madrid. See *Salut les Copains*, n°14, September 1963.

[79] The last letter published in the consulted issues of the magazine is particularly revealing: 'Rather sad news for you, Daniel: I'm leaving my province and I'm settling in Paris. All your phony stars, I will now have the pleasure to boo them on site. Regarding your rubbish paper, I won't even need to buy it anymore. I'll likely find it in the nearest bin (they're not lacking them in Paris). You're warned. I'm keeping watch' *Salut les Copains*, n°17, December 1963.

[80] Two of such letters can be found in *Salut les Copains*, n°11, June 1963.

[81] Eliseo Veron, 'L'Analyse du contrat de lecture: une nouvelle méthode pour les études de positionnement des supports presse', in Institut de Recherche et d'Etudes Publicitaires (IREP), *Les Médias: expériences et recherches actuelles* (Paris: IREP, 1985), pp. 203–230.

[82] Charon, *La Presse Magazine*, p. 5.

doing with its listeners.[83] The letters also acted as intertextual and intermedial elements, inferring the strong interrelations between the magazine and other media of popular culture.

One section of the *Salut les Copains* magazine appears of particular interest in regards to the inclusion of readers in its pages. From the third issue onwards, in October 1962, the magazine included a pen pal section, in which short profiles of readers were published. Those profiles included the name, age, gender, interests (hobbies, sports, and favourite artists), and personal address of those interested in receiving letters from other *copains*. According to his editorial in the third issue, Filipacchi started this new section based on suggestions he received from readers[84]: 'Hundreds and hundreds of letters are piling up on my desk: mates are writing to me from all around France. [...] However, after reading this mail, some new sections are mandatory, for instance, the section for pen pals which we inaugurate in this issue (I'm looking for a mate, page 32), which will be a friendship chain, not only in France, but throughout the world'.[85] As an opening to this new section in the magazine, a short text, presented as extracted from a letter from a reader, states: 'We are thousands of boys and girls who are listening every day the same show, who are reading every month the same magazine, who have the same tastes and Daniel as a mate, but who are living without knowing one another'.[86] Such a description of the readers-listeners of *Salut les Copains* echoes the definition of *imagined communities* by Benedict Anderson,[87] suggesting some reader-listeners had a sense of this community.

The short portraits published in the magazine offer particularly interesting insights into its readership and how it was used to display a specific image of *Salut les Copains*, which also echoed the station's image. More specifically, the portraits gave information about some readers, although in a limited and biased way,[88] and inform on their wish to be connected to

[83] The use of letters was not only done by commercial stations. For instance, the BBC conducted, during its first years, research on its audience thanks to unsolicited letters sent by listeners. Andrew Crisell, *Understanding Radio* (London: Routledge, 1994, 2nd ed.), p. 205.

[84] Therefore reinforcing the display of the aforementioned *contrat de lecture*.

[85] *Salut les Copains*, n°3, October 1962.

[86] *Salut les Copains*, n°3, October 1962.

[87] See Anderson, *Imagined Communities*.

[88] Biased in the sense that not all portraits sent to the magazine were published.

one another. The main point of the next few pages, however, is the international image displayed by the magazine through the portraits. If the majority of the potential pen pals were from France and neighbouring French-speaking countries, a large number were from a wide range of other countries. By displaying such profiles on its pages, the magazine truly displayed the community of *copains* as a diverse[89] and global one, as well as a community transnationally active, as the letters and the magazines circulated across borders. The data used (country of origin and gender of the portraits published, see Fig. 8.1) in the following paragraphs are extracted from 13 issues of *Salut les Copains* which included such pen pal sections. As mentioned, the first time the section was published was in the third issue, in October 1962, and, from then, was included in most issues—although not all—until sometime around 1970, when the section appears to have been discontinued. The number of portraits published in the section changed greatly in the first issue. While the third published only four portraits, the fifth published a total of fifty, and the eleventh, fourteenth, and seventeenth, published around twenty each. After that, the following issues appeared to have been more regular, as they published nine portraits in each section.

In total, 183 portraits were collected. Out of those, there is almost a balance between genders, with only five more female portraits, which echoes the magazine's unisex line.[90] As mentioned, the majority of portraits (53,5%) were from young people living in France, followed by Belgium and Switzerland,[91] meaning that around 65% of the portraits were from the heart of the expected readership, found in French-speaking countries of Western Europe. These figures inform researchers on the community of *copains* displayed as being overwhelmingly made of native French speakers from Western Europe.[92]

This also means that around a third of the *copains* whose portrait was published were from countries which did not belong to this core of the readership, meaning that the magazine still displayed a strong foreign

[89] If some diversity is found in this section, as shown by Mat Pires, the community of *copains* was mostly imagined as white. Pires, 'Les stars noires et *Salut les Copains*', pp. 59–71.

[90] Blandin, 'Radio et magazine: une offre plurimédia', pp. 134–142.

[91] More precisely, from French-speaking parts of those countries.

[92] Only 3 portraits from the French-speaking part of Canada were published in the corpus, making the number too low to be used.

Country of origin	Number of portraits
France	98
Belgium	12
Switzerland	9
West Germany	8
United Kingdom	7
Other Western European countries (Netherlands, Austria, Portugal, Italy, Spain, Greece)	16
Total Western Europe	150
North Africa (Morocco, Algeria, Tunisia)	8
South Vietnam	3
Total former French colonies	11
Eastern bloc (Poland, Romania, Hungary)	5
North America (Québec, United States)	5
Turkey	3
Other countries only mentioned once (Brazil, Yugoslavia, Syria, South Africa, Mauritius)	5
In barracks for military service (in France, Germany and Belgium)	4
Gender	
Male	89
Female	94

Fig. 1 Country of origin and gender of the portraits in the Pen Pal section of *Salut les Copains*

reach. Among these groups, three trends can be discerned. First, there is a strong presence of other Western European countries, with West Germany and the United Kingdom[93] at the top of the list. This is no surprise, considering the geographical closeness and large population of those countries. It also means that the community of *copains* extended to neighbouring countries to this heart of the readership, where it was not impossible for Europe n°1 and *Salut les Copains* to be heard, and where many of the cultural norms, especially in terms of movies and records, were shared. A second group of countries are made of former French colonies. On one hand, eight *copains* from Morocco, Algeria, and Tunisia had their profiles published, while, on the other hand, so did three readers from South Vietnam, creating interesting postcolonial connections. The third trend is found in countries from the Eastern bloc. Although there were only five portraits published from Romania, Poland, and Hungary, it points to the cultural permeability of the Iron Curtain, something already present with the question of radio waves to Eastern Europe.

Overall, the pen pal section of *Salut les Copains* displays an image of its community of readers as both male and female, as well as international. And this internationality operates in a similar way to the transnational space analysed previously. Its core is also focused on France and French-speaking European countries, with similar dynamics of margins and far-away places. The main difference being that some margins, especially with countries in Northern Africa and South Vietnam, are included not so much due to geographical closeness, but cultural and historical connections. Other countries, which only appear on one or two occasions, *e.g.* the United States, Brazil, and Syria, tend to support the idea of *copains* all over the world suggested by Filipacchi in his editorial of October 1962. As mentioned, these pen pal sections cannot be used to inform on the actual readership, although they do indicate that it was a diverse one. It is quite likely that the magazine purposefully promoted these more exotic and unusual portraits, in order to portray itself as a global, transnational actor. By extension, due to the close connections between the radio show, its main figure, Daniel Filipacchi, and to a lesser extent, Europe n°1 also benefited from this image.

The purpose of the pen pal section, according to the magazine, was an answer to readers' queries to be able to exchange with each other, as

[93] More precisely, out of the 7 portraits, 6 were from England and 1 from Northern Ireland.

some felt they belonged to an *imagined community*, although they did not name it as such. The shared culture and interests, as well as the sense of belonging to the same group, due to a shared language—French—and a similar age group.[94] Reading through the profiles, however, reveals two other interests that transpire through the short texts. A potentially romantic one, as an important number of *copains* specifically looked for pen pals of the other gender,[95] while describing themselves as good-looking. It seems that a form of exoticism and curiosity also played a role, as many *copains* from French-speaking countries wrote they were looking to communicate with young people from foreign countries, while those not from this 'core' of the readership often seemed to have an interest in pen pals from Paris. Furthermore, if it is not possible to measure the intensity of letters exchanged on average, one particular profile seems to have attracted strong interest. In the 30th issue, the magazine published a short letter from one of its readers, a woman from Germany named Marion, who thanked 'Daniel' for publishing her address in a previous issue. She also apologises for not being able to reply to all the letters she received, as they were slightly over 1,000. It is extremely likely that this high number of letters received is more the exception than the norm, nevertheless, it reveals a potentially intense exchange of correspondence between the *copains* through the pen pal section. Moreover, despite not being able to measure these exchanges, it is clear that an invisible network of pen pals operated behind the magazine. It was also a transmedial one, as it included issues of the magazines to obtain the addresses, letters as the core medium of communication, as well as a wider shared culture, in which Europe n°1, movies and records, appear to have played a key role.

As this chapter has shown, interactivity and inclusion of readers was an important feature of *Salut les Copains*. Although, while the question of interactivity is stronger with *Salut les Copains*, hence the focus in this chapter, the British magazine also included its readers. It offered them the possibility to write features, fictional or not, and to send photographs of celebrities that would be published in the magazine, in exchange for

[94] Most of the portraits were from young people between 16 and 20 years old at the time.

[95] As shown by Chris Tinker, *Salut les Copains* did not leave much space for non-heteronormativ norms, which could support this hypothesis. Tinker, 'Mixed Masculinities', pp. 84–108.

money.[96] This practice might be a way for the magazine to easily produce content, as it echoes paying journalists at piecework rate, but it is nevertheless a form of inclusion of its readers, as they would have their name, alongside their texts and photos, published in *Fabulous 208*. Moreover, the article inviting readers to submit contributions indicates that rules and deadlines are different depending on their countries, classifying them between readers in the United Kingdom, readers in continental Europe, and readers in countries such as the United States, Australia, and New Zealand. By doing so, this article points to its international readership. Overall, the inclusion of readers in the pages of the magazines was an opportunity for *Fabulous 208* and *Salut les Copains* to give their readership a feeling of being heard, to shape the magazines' international image, and to further the creation of a sense of community, especially for the *copains*; all elements echoing similar dynamics found to some extent in the radio programmes.

References

Amossy, Ruth. *La Présentation de soi. Ethos et identité verbale*. Paris: Presses Universitaires de France, 2010.

Anderson, Benedict. *Imagined Communities: Reflections on the Origin and Spread of Nationalism*, 3rd ed. London: Verso, 2006.

Barthes, Roland. *Mythologies*, revised ed. Translation by Jonathan Cape. London: Vintage, 2009.

Blandin, Claire, ed. *Manuel d'analyse de la presse magazine*. Malakoff: Armand Colin, 2018.

Blandin, Claire. "Radio et magazine : une offre plurimédia pour les jeunes des Sixties". *Le Temps des Médias* 21:2 (2013): 134–142.

Brake, Laurel. "Aestheticism and Decadence: *The Yellow Book* (1894–7), *The Chameleon* (1894), and *the Savoy* (1896)". In *The Oxford Critical and Cultural History of Modernist Magazines. Volume I: Britain and Ireland, 1880–1955*, edited by Andrew Thacker and Peter Brooker, 76–100. Oxford: Oxford University Press, 2013.

Brake, Laurel, and Marysa Demoor, eds. *Dictionary of Nineteenth Century Journalism*. Ghent: Academia Press, 2009.

Charon, Jean-Marie. *La Presse Magazine*. Paris: La Découverte, 2008.

Chignell, Hugh. *Key Concepts in Radio Studies*. London: Sage, 2009.

[96] *Fabulous 208*, 14th January 1967.

Craig, Douglas B. *Fireside Politics: Radio and Political Culture in the United States, 1920–1940*. Baltimore: John Hopkins University Press, 2006.

Crisell, Andrew. *Understanding Radio*, 2nd ed. London: Routledge, 1994.

Cronqvist, Marie and Christoph Hilgert. "Entangled Media Histories. The Value of Transnational and Transmedial Approaches in Media Historiography". *Media History* 23:1 (2017): 130–141.

Dakhlia, Jamil. "Propriétés et fonctions de la presse magazine". In *Manuel d'analyse de la presse magazine*, edited by Claire Blandin, 51–65. Malakoff: Armand Colin, 2018.

Delporte, Christian, Claire Blandin and François Robinet. *Histoire de la presse en France : XXe-XXIe siècles*. Malakoff: Armand Colin, 2018.

Ferrand, Franck. "Au coeur de l'émission Salut les Copains !". *Au Coeur de l'Histoire*. Europe 1. September 14th, 2012.

Fickers, Andreas. "'Neither Good, Nor Bad; Nor Neutral': The Historical *Dispositif* of Communication Technologies". In *Journalism and Technological Change: Historical Perspectives, Contemporary Trends*, edited by Martin Schreiber and Clemens Zimmermann, 30–52. Frankfurt: Campus, 2014.

Fiske, John. *Television Culture*. London, UK: Methuen, 1987.

Fleury, Adeline. "Exposition: les sixties exquises de Jean-Marie Périer en images". *Le Parisien*, January 17th, 2019, [http://www.leparisien.fr/culture-loisirs/exposition-les-sixties-exquises-de-jean-marie-perier-en-images-17-01-2019-7989615.php].

Fourment, Alain. *Histoire de la presse et des journaux d'enfants (1768–1988)*. Paris: Eole, 1987.

Gaudreault, André. *Du littéraire au filmique: Système du récit*. Malakoff: Armand Colin, 1999.

Gripsrud, Jostein. *Understanding Media Culture*. London: Arnold, 2002.

Hendy, David. *Radio in the Global Age*. Cambridge: Polity Press, 2000.

Hilmes, Michele. "Radio and the Imagined Community". In *The Sound Studies Reader*, edited by Jonathan Sterne, 351–362. Abingdon-on-Thames: Routledge, 2012.

Horwitz, Robert. *The Irony of Regulatory Reform: The Deregulations of American Telecommunications*. Oxford: Oxford University Press, 1989.

Jensen, Klaus Bruhn. "Intermediality". In *The International Encyclopedia of Communication Theory and Philosophy*, edited by Klaus Bruhn Jensen et al., 972–982. Chichester: John Wiley & Sons, 2016.

Jobling, Paul and Richard Crowley. *Graphic Design: Reproduction and Representation since 1800*. Manchester: Manchester University Press, 2003.

Legay, Richard. "Studying 1960s Commercial Radio". In *Bloomsbury Handbook of Radio*, edited by McDonald, Kathryn and Hugh Chignell, 367–382. London: Bloomsbury, 2023.

Looseley, David. "In from the Margins: *Chansons*, Pop and Cultural Legitimacy". In *Popular Music in France from Chanson to Techno*, edited by Hugh Dauncey & Steve Cannon, 27–40. Aldershot: Ashgate, 2003.

Maldener, Aline. "Fabulous consumerism? Mediale Repräsentationen jugendlicher Konsumkultur in westdeutschen, britischen und französischen Jugendzeitschriften der 1960er und 1970er Jahre". In *Populärkultur transnational: Lesen, Hören, Sehen, Leben im Europa der langen 1960er Jahre*, edited by Dietmar Hüsern 199–224. Bielefeld: Transcript, 2017.

Martinez Alfaro, Maria Jesus. "Intertextuality: Origins and Developments of the Concept". *Atlantis* 18:1–2 (1996): 268–285

Méchoulan, Éric. "Intermédialités : le temps des illusions perdues", *Intermédialités /Intermediality* 1 (2003): 9–27.

Müller, Jürgen E. "L'intermédialité, une nouvelle approche interdisciplinaire : perspectives théoriques et pratiques à l'exemple de la vision de la télévision". *Cinémas* 10:2–3 (2000): 105–134.

Périer, Jean-Marie. *Oncle Dan. Souvenirs*. Paris: XO Éditions, 2008.

Pires, Mat. "The Popular Music Press". In *Popular Music in France from Chanson to Techno*, edited by Hugh Dauncey & Steve Cannon, 77–96. Aldershot: Ashgate, 2003.

Pires, Mat. "Les stars noires et *Salut les Copains*, 1962–1968". *Communication et langages* 111 (1997): 59–71.

Prior-Miller, Marcia R. "Magazine Typology. Using Operational Classification Technology". In *The Routledge Handbook of Magazine Research. The Future of the Magazine Form*, edited by David Abrahamson, Marcia R. Prior-Miller, 22–50. Abingdon: Routledge, 2015.

Provenzano, François. "Le Discours de la presse magazine". In *Manuel d'analyse de la presse magazine*, edited by Claire Blandin, 85–104. Malakoff: Armand Colin, 2018.

Quillien, Christophe. *Nos Années Salut les Copains 1959–1976* (Paris: Flammarion, 2009) & Gilbert Jouin, *Salut les Copains*. Paris: Ipanema Éditions, 2012.

Rajewski, Irina. *Intermediales Erzählen in der italienischen Literatur der Postmoderne von den giovani scrittori der 80er zum pulp der 90er Jahre*. Tübingen: Gunter Narr, 2001.

Rajewski, Irina. "Intermediality, Intertextuality, and Remediation: A Literary Perspective on Intermediality". *Intermédialités /Intermediality* 6 (2005): 43–64.

Remoortel, Marianne Van, Kristin Ewins, Maaike Koffeman and Matthew Philpotts, "Joining Forces: European Periodical Studies as a New Research Field". *Journal of European Periodical Studies* 1:1 (2016): 1–3.

Rixon, Paul. “The Role of British newspapers in framing the public perception and experience of European radio, 1930–1939”. *Media History* 20:2 (2020): 153–166.

Seegers, Lu. *Hör zu!: Eduard Rhein und die Rundfunkprogrammzeitschriften (1931–1965)*, 2nd ed. Potsdam: Verlag für Berlin-Brandenburg, 2003.

Stamm, Michael. *Sound Business: Newspapers, Radio, and the Politics of New Media*, 2nd ed. Philadelphia: University of Pennsylvania Press, 2016.

Stamm, Michael. “The Sound of Print: Newspapers and the Public Promotion of Early Radio Broadcasting in the United States”. In *Sound in the Age of Mechanical Reproduction*, edited by David Suisman and Susan Strasser, 221–241. Philadelphia: University of Pennsylvania Press, 2010.

Thomson, Olivia. “The History of Radio Times”. *Radio Times*, July 17th, 2013. https://www.radiotimes.com/news/2013-07-17/the-history-of-radio-times/

Tinker, Chris. “Mixed Masculinities in 1960s British and French Youth Magazines”. *The Journal of Popular Culture* 7:1 (2014): 84–108.

Tinker, Chris. “Shaping 1960s youth in Britain and France: Fabulous and Salut les Copains”. *International Journal of Cultural Studies* 14:6 (2011): 641–657.

Trachtenberg, Alan. *The Incorporation of America: Culture and Society in the Gilded Age*. New York: Hill and Wang, 1982.

Veron, Eliseo. “L’Analyse du contrat de lecture : une nouvelle méthode pour les études de positionnement des supports presse”. In *Les Médias : expériences et recherches actuelles*, edited by Institut de Recherche et d’Etudes Publicitaires (IREP), 203–230. Paris: IREP, 1985.

CHAPTER 9

Highly Visual Culture and Stardom

Looking at the cultural content produced by commercial radio stations and their partners reveals how visual this content could be. *Fabulous 208* and *Salut les Copains* were periodicals designed to accompany and extend the programmes of Radio Luxembourg and Europe n°1, and were filled with photos of celebrities, which were part of their popularity,[1] and included, at times, posters that could be cut out of the magazines, furthering a sense of appropriation of this popular culture.[2] Magazines like those mentioned above played a key role in the development of stardom, which was furthered by this visual culture, as well as commercial radios, which were crucial actors to promote not only the records but the artists beyond them. Moreover, the stardom culture present in the magazines, which echoed what happened on air, also promoted radio hosts, especially disc-jockeys, as part of this culture, and in ways not dissimilar to musical artists and movie stars. The creation of celebrities by Hollywood described by Richard Dyer shows the relevance of including the question of stardom in this chapter. Defined as commodities manufactured by the media industries, stars belong to a 'complex image-making system'

[1] Chris Tinker, "Shaping 1960s Youth in Britain and France: Fabulous and *Salut les Copains*", *International Journal of Cultural Studies* (14:6, 2011), pp. 641–657.

[2] David Looseley, "Conceptualising Youth Culture in Postwar France", *Modern & Contemporary France* (15:3, 2007), pp. 261–275.

R. Legay, *The Cultural Histories of Radio Luxembourg and Europe n°1*,
https://doi.org/10.1007/978-3-031-46250-4_9

that operate on a multimedia and intertextual level,[3] and the magazines studied in this chapter directly contributed to this phenomenon, notably by exploiting their own strengths in the media landscape.

Photos and images have always been a key specificity of magazines, from their very beginning in the nineteenth century, and a way for them to stand out as a new genre within the larger field of press.[4] This led magazines to always try to be at the cutting edge of printing technology and to work with paper of higher quality than in the daily press, as photos played a key role in shaping the identity of a magazine.[5] And, it can be argued, in the case of the periodicals studied in this chapter that it extended to the identity of radio shows and radio stations. The importance of photography for *Salut les Copains* was mentioned in the previous chapter, through the work of Jean-Marie Périer, although he was not the only photographer working for *Salut les Copains*.[6] He nevertheless held a supervising position over other photographers, and, reflecting on his collaboration with the magazine, he explained he had a wonderful position. *Salut les Copains* was a very profitable business, therefore he had no logistical and financial issues, giving him *carte blanche* for his work—including last-minute trips to faraway locations[7]—indicating the importance of photos for the magazine. As mentioned by Barthes, '*la photo de presse*', is an object 'well-made, chosen, composed, constructed, treated according to professional, aesthetic or ideological norms',[8] supporting further the importance of studying them with great detail and attention.[9]

[3] Richard Dyer, *Heavenly Bodies: Films Stars and Society* (Abingdon: Routledge, 2004, 1st ed. 1996), pp. 3–5.

[4] Jean-Marie Charon & Rémi Rieffel, "La Presse magazine", *Réseaux* (105, 2001), pp. 9–16.

[5] Jamil Dakhlia, "Propriétés et fonctions de la presse magazine", in Claire Blandin (ed.), *Manuel d'analyse de la presse magazine* (Malakoff: Armand Colin, 2018), pp. 51–65.

[6] Tony Frank and Jean-Jacques Damour were other well-known photographers of the magazine.

[7] This partly explains the vast number of photo-reportages to foreign locations. See Gilbert Jouin, *Salut les Copains* (Paris: Ipanema, 2012), p. 51.

[8] Roland Barthes, "Le message photographique", *Communication* (1, 1961), pp. 127–138.

[9] See, for instance, Annette Vowinckel, "Das Bild als Ding zwischen der zweiten und der dritten Dimension", in Andreas Ludwig (ed.), *Zeitgeschichte der Dinge: Spurensuchen in der materiellen Kultur der DDR* (Göttingen: Vandenhoeck & Ruprecht, 2019), pp. 65–84.

1 Omnipresence of Photos in the Stardom Culture

A clear purpose of *Salut les Copains* and *Fabulous 208* was to contribute to a general stardom culture,[10] by promoting a series of artists, mostly from the music industry, although movie stars often appeared in the pages of the periodicals.[11] As shown by Dakhlia, magazines, unlike newspapers, cannot cover current affairs, notably due to their publication rhythm, which partly explains their regular engagement with stardom culture.[12] More precisely, magazines have often engaged with journalism focused on the 'human', embracing a personal and affective angle. Even a brief look at issues of either magazine studied in this chapter reveals that pages and photo-reportages dedicated to musicians and singers have little to do with their music. And a lot with their personal lives,[13] supporting the idea that they fully subscribe to this magazine practice. Among the 'stars' promoted by *Salut les Copains* and *Fabulous 208*, listeners could read about well-known musicians, such as, respectively, Johnny Hallyday[14] and The Beatles for instance, as well as lesser-known ones. In the pages of both

[10] The global phenomenon of stardom has been studied at length, and in many Western European countries. See John Gaffney & Diana Holmes, *Stardom in Postwar France* (New York: Berghahn Books, 2011); Stefanie Herrmann, *50 Jahre BRAVO Eine Jugendzeitschrift als Spiegel der Zeitgeschichte* (Berlin: Archiv der Jugendkultur Verlag, 2014); Sean Redmond, *Stardom and Celebrity: A Reader* (Los Angeles: Sage, 2010).

[11] In an article published in 1967, Leroy and Quesnel studied the cover pages of 14 French magazines to see patterns and tendencies regarding what they called 'opinion leaders'. As part of their study, they established that *Salut les Copains*—as well as *Mademoiselle Âge Tendre*, another periodical published by Filipacchi—were engaging in a 'cult of personalities'. See Alain Leroy & Louis Quesnel, "Un indice objectif de leadership: Les photographies de couverture des magazines", *Les Cahiers de la publicité* (n°18, 1967), pp. 108–110.

[12] Dakhlia, "Propriétés et fonctions de la presse magazine", in Blandin, *Manuel d'analyse de la presse magazine*, pp. 51–65.

[13] For instance, *Salut les Copains* published a two-page article asking French actor Alain Delon his thoughts—among other things—on wearing shorts, why he sold his car, and the last witty comment from his son. *Salut les Copains*, n°103, March 1971. *Fabulous 208* published a similar article in which British singer Dave Lee explained in detail how difficult it is to have friends when 'being in the show-biz'. *Fabulous 208*, 5th June 1971.

[14] The *yéyé* singer was indeed on the cover of *Salut les Copains* on countless occasions, and the magazine dedicated a special supplement to his wedding with Sylvie Vartan in April 1965, see *Salut les Copains*, n°34, special supplement, April 1965. He also held a prominent position in the '*photo du siècle*' mentioned previously.

magazines, many artists were—figuratively—rubbing shoulders with each other. Photo-reportages appear to have been the main way to promote specific figures, as the format mixed high-quality pictures with texts, therefore playing on the specificities of the magazine.[15] One interesting example is found in an issue of *Salut les Copains* from 1966, in which a few pages are dedicated to seven young singers, thought, by the editorial team, as upcoming.[16] Their photos were of high quality, similar to those taken for well-known celebrities, however, the texts accompanying them echoed those presenting readers in the pen pal section, focusing on their (young) age, physical features, and hobbies. As such, they appeared as a group of people halfway between the celebrities promoted in *Salut les Copains* and the readers of the magazine.

Overall, the magazines developed a scattering of personalities, who were therefore linked to the periodicals, and, by extension, to the radio stations. Indeed, Radio Luxembourg and Europe n°1 were regularly playing the records of the artists whose pictures were found in the magazines, and they were regularly invited in the studios, and for other events organised by the stations. This—the entanglements between commercial radio stations, magazines, and specific figures of the music industry—is a first aspect through which *Fabulous 208* and *Salut les Copains* differentiated themselves from other similar publications of the Long Sixties. The second one is how they included some of their 'own' figures into the stardom. The importance of a cult of personality surrounding some radio hosts and disc-jockeys was already mentioned when talking about the sound of commercial radio, however, this element is even more striking in their printed publications.[17] Both in *Fabulous 208* and *Salut les Copains*, radio personalities from commercial stations rubbed shoulders with artists and other celebrities, putting them—and, by extension, the stations—in the same imaginary footing as well-known figures of the entertainment

[15] Photo-reportages about artists were common in both magazines, and they covered a wide range of artists such as British rock band The Rolling Stones, French singer-songwriter Georges Brassens, and even French actor Alain Delon, to quote a few examples.

[16] *Salut les Copains*, n°46, May 1966.

[17] It seems important to note that radio personalities already had celebrity status in the 1920s and 1930s. See, for instance, *Das Buch der Ansager: Die Ständigen Rundfunkansager der Europäischen und der Grossen Amerikanischen Rundfunkgesellschaften in Wort und Bild* (Berlin: Rothgiesser & Diesing A.G., 1932), which portrays many radio hosts of the time.

and music industries. If both magazines applied this technique of visually representing radio personalities as members of a wider stardom, they did it with significant differences.

In the pages of *Salut les Copains*, Daniel Filipacchi had a central role. His name was regularly mentioned, letters sent by readers were addressed to him, and he sometimes wrote editorials and other short pieces, usually addressing readers directly. He was also portrayed on a few occasions in *Salut les Copains*, in which he held an important dual figure. On one hand, he was another *copain*, as were his readers—and listeners—whom he addressed as friends, while always signing by his first name.[18] On the other hand, he also held a more paternalistic position, as he was significantly older than his readership—and audience—and the person in charge of the radio show and the magazine.

The portrayal of 'Daniel' in *Salut les Copains* is often an occasion to depict him as a peer to other 'established' celebrities. One example of that is found in the fifth issue, from December 1962.[19] A photo-reportage pictures him, on a double-page, with French singers Sylvie Vartan, Françoise Hardy, and Johnny Hallyday, discussing in the studios of Europe n°1. The four of them appear on the same line, and portrayed as equals. A few pages later in this reportage, Filipacchi is shown next to other artists, such as Petula Clark and Fats Domino. It is important to remember that he is not portrayed as a journalist interviewing the artists, but rather as one of them. All of the artists mentioned above were regularly portrayed in *Salut les Copains*, therefore, having Filipacchi among them, as a peer, operated as a form of validation of his status as a celebrity like they were.

Fabulous 208, on the other hand, offered a more diverse image, representing various disc-jockeys and hosts in its pages, which is consistent with the magazine's links with a station, rather than a specific show. Disc-jockeys of the station were treated and portrayed in a similar way to other celebrities.[20] This included, for instance, articles and interviews

[18] A letter written by 'your friend, Daniel', with a portrait of Filipacchi, can be found in *Salut les Copains*, n°3, October 1962.

[19] *Salut les Copains*, n°5, December 1962.

[20] In this regard, the example of Vera Lynn (1917–2020), the popular British singer and entertainer, points at the long history of radio personalities and stardom. See Christina Baade, '"Sincerely Yours, Vera Lynn": Performing Class, Sentiment, and Femininity in the

detailing their personal lives,[21] and even holiday plans.[22] The celebrity culture surrounding radio personalities also made use of magazines' peculiar relationship to images, and, as such, portraits of Radio Luxembourg disc-jockeys were common in the pages of *Fabulous 208*. Apart from Jimmy Savile's portrait on the cover on one occasion, as mentioned in the previous chapter, disc-jockeys were never again on the front pages of *Fabulous 208*. They were, nevertheless, at the centre of many photo-reportages[23] and other photoshoots published in the magazine.

2 The *A-Z* of *Fab Fancies*

In June 1967, Fleetway published the *A-Z of Fab Fancies*[24]; a book containing details and pictures of 400 celebrities, mostly from the United Kingdom.[25] The *A-Z* appears to have been published as a special issue of *Fabulous 208*, however, readers had to detach, cut, and fold in the

"People's War"', *Atlantis* (30:2, 2006), pp. 36–49, and Paddy Scannell, *Radio, Television and Modern Life: A Phenomenological Approach* (Oxford: Blackwell, 1996), pp. 58–74.

[21] For instance, in 1968, a short article was published in *Fabulous 208*, in which the journalist Julie Webb described her visit to Radio Luxembourg's disc-jockey Tony Blackburn's flat in London, and the questions she asked him about it. The reader could learn in great detail the disc-jockey's taste in furniture and cooking, as well as his habit to bring his dirty laundry back to his mum in Bournemouth. *Fabulous 208*, 8th June 1968.

Another example can be found in an interview with the American disc-jockey Rosko, who tells anecdotes about his personal life and hobbies, with a close-up portrait in the middle of the page. Interestingly, a similar article is published a few pages later, this time with the Scottish singer-songwriter singer Donovan. *Fabulous 208*, 10th January 1970.

[22] A double-page article published in 1971 details, for instance, the summer holiday plans of ten disc-jockeys of Radio Luxembourg (among them: Tony Blackburn, Kid Jensen and Noel Edmonds). *Fabulous 208*, 5th June 1971.

[23] The example of Prince, Jensen, and Burnett photographed during a horse-riding lesson (*Fabulous 208*, 12th July 1969) mentioned previously in this chapter represents a good example. A more 'serious' example is found in an issue from 1967, in which close-up portraits of five Radio Luxembourg disc-jockeys (Dave Cash, Stuart Grundy, Don Wardell, Jimmy Savile, and Cathy McGowan) were published, alongside a short text with their names and current activities. *Fabulous 208*, 3rd June 1967.

[24] Henceforth abbreviated as *A-Z*.

[25] June Southworth (ed.), *A-Z of Fab Fancies. Details of over 400 Top Names* (London: Fleetway, 1967), 128 pages.

middle of the pages of the issue to make it into the final product.[26] The *A-Z* is a pocket-sized book, made of rather cheap paper, with soft covers, in colour, while most of the pages are in black and white.[27] 400 names of celebrities (as individual artists, or as bands) are listed, in alphabetical order, offering a small picture, some biographical and anecdotal details, as well as an address to write to them. In most cases, each page details five different names; however, some celebrities are offered a larger space,[28] creating de facto a hierarchy between them. Personalities—or 'fancies' as they are called—listed in the book are mostly musicians and songwriters, although there are a few actors as well.[29] There is a strong presence of male figures, and mostly from the United Kingdom, then from the United States. A few celebrities from other countries, notably the French-speaking world are listed as well,[30] showing the presence, although discreet, of a global popular culture in a book destined to a British audience. In the letter of the editor, one can also find a hint at a cross-generational index of celebrities: 'We've tried to find facts on all your "fancies" and on some of your Mum's, too', by the inclusion of some names that would be seen as more relevant to an older generation than those of the targeted readership.

What is of particular interest here, however, is the presence of a handful of radio personalities among the musicians and actors. Out of

[26] Which echoes the concept of 'creative consumer', used for instance by Looseley to explain how artefacts of mass culture are appropriated by consumers through re-work, *bricolage*, in creative and dynamic ways. See Looseley, 'Conceptualising Youth Culture', pp. 261–275.

[27] With the exception of some full-page pictures which are in colour, such as a portrait of Swiss actress Ursula Andress (born in 1936), known for her role in *Dr. No* (1962) holding a black cat on page 9, therefore imitating the usual style of *Fabulous 208*.

[28] For instance, the Beach Boys are on a full-page: the top half being a picture of the band, while the bottom half a longer text introducing them. The Beatles are presented over two pages; one for the band, the other for the individual members.

[29] Ursula Andress, already mentioned, but also, among others, David McCallum, Michael Caine, and Ian McShane.

[30] Françoise Hardy, Alain Delon, Claude François, and Johnny Hallyday are part of this 'A-Z' for instance. Indian musician Ravi Shankar and Egyptian actor Omar Sharif are two other examples of non-British 'fancies'.

the 400 names, 13 can be defined primarily as radio hosts and disc-jockeys.[31] All of them worked at the English service Radio Luxembourg, except one—Mike Lennox—who was the 'star DJ' of the newly launched BBC Radio 1. This reveals an obvious bias in favour of the commercial radio station, which was working alongside the magazine *Fabulous 208* and its publisher. Two biographies (those of Dave Cash and Don Wardell) even refer to the collaboration between the disc-jockeys and the magazine, in which they published columns, inferring clear intermedial references. The portraits of the disc-jockeys all followed a similar pattern, including their age or date of birth, their physical appearance (hair and eye colour, height and weight), one or two positive comments regarding their work on radio, where to contact them,[32] and, quite interestingly, a short 'backstory' of how they ended up on 208. For two of them—Tony Blackburn and Simon Dee—their biographies mention their time on pirate stations, and for others—Simon Dee, David Jacobs, and Don Wardell -, they mention their time in the army, more precisely in the British Forces Network. Overall, the format used for the 13 radio hosts is not so different from the one used for other 'fancies', therefore including them fully as part of the wider pool of celebrities presented by the book. As almost all of them were actually working on Radio Luxembourg,[33] the inclusion of their names in the book can be seen as an enterprise of legitimisation of the figure of the radio host as part of the wider celebrity culture, putting them on the same level as music idols and movie stars. By extension, the source presents *Fabulous 208* and Radio Luxembourg as major actors and media of popular culture, alongside record labels and studios.

One small section of the source echoes this idea of being at the crossroad of multiple media. The four last pages of the book are dedicated to addresses and contact points of key actors in four different fields. The

[31] These 13 men were picked because their radio activities were central, in opposition to other 'fancies', such as actors and TV hosts, for whom radio was only part of a larger portfolio. They were 'The Baron', Tony Blackburn, Pete Brady, Dave Cash, Chris Denning, Simon Dee, Kenny Everett, Alan Freeman, David Jacobs, Mike Lennox, Peter Murray, Jimmy Savile, and Don Wardell.

[32] Most of the time, at 38, Hertford Street, London, the headquarters of Radio Luxembourg in Britain.

[33] It could easily be argued that many other names of famous radio personalities, who were not working for 208, but were just as famous, could have been included as well, insisting on the clear choice to promote the station.

source lists, for instance, the main record and film companies in Britain, but also where to go, around London, to work as an extra on films. A similar page lists British television stations, and one page focuses on radio. Interestingly, the page mostly deals with the new BBC radio stations, and where to write for 'request programmes' and to attend shows with an audience. The page also details the new upcoming local radio stations, giving only a few lines to Radio Luxembourg and its address on Hertford Street. Through these four pages, the source operates as a reference point for four different media of popular culture (radio, television, film, and music), bringing them together to inform its readers. The magazine and the radio station bring themselves at the crossroads of popular youth media through this book, and in an overseeing position, rather than a mere actor among others.

The purpose of the book itself deserves to be analysed as well. According to the editorial, the team behind *Fabulous 208* was so overwhelmed by letters asking for details about celebrities, that they decided to publish one special issue which would cover most of the inquiries received. The end product is then a compact book identifying 400 key names of popular culture, although without clear justifications, portraying itself as a reference point at a specific time. A commercial purpose cannot be neglected, as such supplements were—and still are—often used to support the sales of youth magazines, which benefit from the extra appeal offered thanks to this merchandising.[34] The need for the reader to cut and fold the magazine to obtain the book suggests a form of appropriation of the material,[35] while the soft paper and the pocket-size of the *A-Z* points at an easy book to carry and share,[36] both elements supporting the idea that magazines, and their supplements, allowed a more creative consumption than the simple action of reading.

[34] Françoise Hache-Bissette, "La Presse pour la jeunesse: entre éducation et recreation", in Blandin, *Manuel d'analyse de la presse magazine*, pp. 203–211.

[35] Other special issue books published with the station, such as almanacs, included sections inviting readers to write personal information (*i.e.* name, address, hair and eye colour) and comments about their favourite artists and records.

[36] The size of a magazine impacts its uses. A large format (A4 and above) points at a 'high-end' publication, destined to 'decorate nightstands and waiting rooms'. Medium and smaller formats (30 × 23 cm and below) are designed to be easily transported. Dakhlia, 'Propriétés et fonctions de la presse magazine', in Blandin, *Manuel d'analyse de la presse magazine*, pp. 51–65.

3 The Creative Consumer

Magazines, according to Jamil Dakhlia, are, thanks to their paper of good quality, their page setting, and their illustrations, 'beautiful objects', which might be compelling for readers to store and collect them,[37] something also encouraged by the magazines themselves.[38] They included and advertised photos to be cut out and turned into posters.[39] Facing such objects of popular culture, readers could engage in an active consumption of the product. The notion of 'creative consumer' suggests that 'artefacts of mass culture, supposedly imposed, reified and productive only of passivity, are in fact re-worked, *bricolés* by their users in ways that are creative and dynamic',[40] and is particularly meaningful here.[41] Indeed, the nature of magazines supported practices of *thésaurisation*[42] (hoarding) of entire issues or specific pages/sections. Mostly used in the context of youth culture studies, the concept of 'creative consumer' is especially deployed to understand popular culture from a reception perspective, which is not the focus of this book. Nevertheless, it is a powerful conceptual tool to shed light on how the magazines studied in this chapter supported and even encouraged such a creative consumption.

Moreover, high-quality photos could be removed to be used as posters, serials published in the magazines could be cut out from successive issues to be assembled, and the example shown above of the *A-Z of Fab Fancies* points to the tinkering encouraged in some special issues. It is indeed

[37] Dakhlia, "Propriétés et fonctions de la presse magazine", in Blandin, *Manuel d'analyse de la presse magazine*, pp. 51–65.

[38] *Salut les Copains* sold 'luxurious bindings' to readers who wanted to store and display their issues of the magazine. See *Salut les Copains*, n° 46, May 1966.

[39] Many issues of *Fabulous 208* advertised their 'pin-ups' and 'king size colours' on their cover page.

[40] Looseley, "Conceptualising Youth Culture", pp. 261–275.

[41] This concept echoes the research undertaken by the Centre for Contemporary Cultural Studies at the University of Birmingham in the 1970s and 1980s, notably through Stuart Hall, which supported an understanding of popular culture as a socially constructed system of codes and signifiers. This perspective gave the consumer—or listener-reader—a more active role, which contrasted with a vision of popular culture as a more subordinating force. See, for instance, Stuart Hall, "Encoding/decoding", in S. Hall, D. Hobson, A. Lowe & P. Willis (eds.), *Culture, Media, Language: Working Papers in Cultural Studies 1972-79* (London: Routledge, 2005, 1st edition 1980), pp. 117–127.

[42] Dakhlia, "Propriétés et fonctions de la presse magazine", in Blandin, *Manuel d'analyse de la presse magazine*, pp. 51–65.

a clear example of a product that was made to be—to use Looseley's words—'re-worked' and '*bricolés*'[43] (tinkered with), as it was cut, folded, and the pages reorganised to be turned into a new product. Moreover, it was designed for this very purpose, showing the editors encouraging their readers for creative consumption. *Salut les Copains* also engaged in these practices, which were certainly not limited to *Fabulous 208*. In a few issues, pages of a '*trombinoscope*[44] of celebrities' were published,[45] encouraging readers to remove them and create a new object from them. In a letter published in 1966, a reader appears delighted that the magazine started to publish a 'lexicon' of cinema, which will be a nice addition to their copy of the aforementioned '*trombinoscope*' and of a 'dictionary of *copains*',[46] which had already been 'cut and bound'.[47] Some of the physical issues of the magazines obtained to conduct the research kept traces of this creative consumption, as indicated by the occasional missing page, cut-out articles and photos, and written comments. These comments might give the name, age and, sometimes, the addresses of a previous owner, who wrote this information likely to assert their ownership of the magazine. However, some written indications are even more interesting here. Both magazines had sections inviting readers to write in them, especially regarding their choices of favourite artists, often asking to rank them, and, at times, previous owners of the magazines did such a thing. Although it would be fascinating to conduct research based on these elements, sources are too scarce and incomplete to do so. Nevertheless, they show the magazines offered the possibility to do so, and encouraged readers to write in and appropriate them, and that some of them did this. Interestingly, the idea of a creative consumer is not incompatible with radio programmes, joining the two worlds. Although far less easy than cutting out a page out of a magazine, radio programmes could be recorded, edited, shared, and copied. This is rather evident as many audio sources consulted in preparation for this book were made by

[43] Looseley, "Conceptualising Youth Culture", pp. 261–275.

[44] A *trombinoscope* is a French word which refers to a document gathering portraits of people, and is often used in organisations and schools.

[45] *Salut les Copains*, n°46, May 1966; and *Salut les Copains*, n°51, October 1966.

[46] The 'dictionary of *copains*' was offering definitions of French slang words of the 1960s. A rather welcome addition for contemporary readers of the magazine, which made regular use of such words.

[47] *Salut les Copains*, n°51, October 1966.

amateurs, suggesting a similar creative consumption and appropriation. Nevertheless, radio stations did not encourage such behaviour, contrary to magazines.

References

Baade, Christina. "'Sincerely Yours, Vera Lynn': Performing Class, Sentiment, and Femininity in the 'People's War'". *Atlantis* 30:2 (2006): 36–49.

Barthes, Roland. "Le message photographique". *Communication* 1 (1961): 127–138.

Charon, Jean-Marie, and Rémi Rieffel. "La Presse magazine". *Réseaux* 105 (2001): 9–16.

Dakhlia, Jamil. "Propriétés et fonctions de la presse magazine". In *Manuel d'analyse de la presse magazine*, edited by Claire Blandin, 51–65. Malakoff: Armand Colin, 2018.

Das Buch der Ansager: Die Ständigen Rundfunkansager der Europäischen und der Grossen Amerikanischen Rundfunkgesellschaften in Wort und Bild (Berlin: Rothgiesser & Diesing A.G., 1932).

Dyer, Richard. *Heavenly Bodies: Films Stars and Society*. Abingdon: Routledge, 2004, 1st ed. 1996.

Gaffney, John and Diana Holmes. *Stardom in Postwar France*. New York: Berghahn Books, 2011.

Hache-Bissette, Françoise. "La Presse pour la jeunesse: entre éducation et récréation". In *Manuel d'analyse de la presse magazine*, edited by Claire Blandin, 203–211. Malakoff: Armand Colin, 2018.

Hall, Stuart. "Encoding/decoding". In *Culture, Media, Language: Working Papers in Cultural Studies 1972–79*, edited by Stuart Hall, Dorothy Hobson, Andrew Lowe & Paul Willis, 117–127. London: Taylor & Francis e-Library, 2005.

Herrmann, Stefanie. *50 Jahre BRAVO Eine Jugendzeitschrift als Spiegel der Zeitgeschichte*. Berlin: Archiv der Jugendkultur Verlag, 2014.

Jouin, Gilbert. *Salut les Copains*. Paris: Ipanema, 2012.

Leroy, Alain and Louis Quesnel. "Un indice objectif de leadership: Les photographies de couverture des magazines". *Les Cahiers de la publicité* 18 (1967): 108–110.

Looseley, David. "Conceptualising Youth Culture in Postwar France". *Modern & Contemporary France* 15:3 (2007): 261–275.

Redmond, Sean. *Stardom and Celebrity: A Reader*. Los Angeles: Sage, 2010.

Scannell, Paddy. *Radio, Television and Modern Life: A Phenomenological Approach*. Oxford: Blackwell, 1996.

Southworth, June, ed. *A-Z of Fab Fancies. Details of Over 400 Top Names*. London: Fleetway, 1967.

Tinker, Chris. “Shaping 1960s Youth in Britain and France: Fabulous and Salut les Copains”. *International Journal of Cultural Studies* 14:6 (2011): 641–657.

Vowinckel, Annette. “Das Bild als Ding zwischen der zweiten und der dritten Dimension”. In *Zeitgeschichte der Dinge: Spurensuchen in der materiellen Kultur der DDR*, edited by Andreas Ludwig, 65–84. Göttingen: Vandenhoeck & Ruprecht, 2019.

CHAPTER 10

Feu de Camp, at the Crossroad Between Radio and Comics

This chapter deals with a specific case study: the short-lived radio programme *Le Feu de camp du dimanche matin*[1] (Sunday morning campfire), in order to analyse the entanglements between radio and comics, more precisely, the commercial radio station Europe n°1 and the comic magazine *Pilote*.[2] *Feu de Camp* was a short-lived radio show, broadcast on French-speaking commercial station Europe n°1 in 1969. It was hosted by creators from *Pilote*, a French comic magazine, thereby creating tangible entanglements between the two fields of comics and radio. Comics and radio have a long history together, for instance, with the radio adaptations of *Tintin*, whose adventures were broadcast on the *Radiodiffusion-télévision française*, the French public broadcaster, or, in the American context, radio dramas of the Dick Tracy and Superman comic series. However, *Feu de Camp* stands out as it aimed to transpose the *spirit* of the irreverent magazine into radio, while other examples would directly adapt comic stories into audio dramatisations.

[1] Henceforth abbreviated as *Feu de Camp*.

[2] With the precious help of comics scholar and historian Jessica Burton from the University of Luxembourg. See Richard Legay & Jessica Burton, "From the Comic Strips to the Airwaves. The Short-lived experiment of Le Feu de Camp du Dimanche Matin on Europe n°1", *Media History* (29:2, 2023), pp. 226–239, for more details.

R. Legay, *The Cultural Histories of Radio Luxembourg and Europe n°1*,
https://doi.org/10.1007/978-3-031-46250-4_10

Both the magazine and Europe n°1 were major actors of popular culture in the French-speaking world in the late 1960s, and as such, this particular programme acts as a gateway into the connections and transfers between some of the most popular media at the time. Nevertheless, there seems to be a lack of public recognition despite the popularity of the magazine and of the station. The uniqueness of the show, coupled with its short-lived nature (having only lasted for 13 episodes), makes it a fascinating case study to consider the links between the two different media. Indeed, only two episodes of the show have been preserved, both of which, never studied before, are considered in this article, in addition to issues of *Pilote* magazine. In view of the nature of archival material available, the concept of entangled media histories, as developed by Cronqvist and Hilgert, is very useful in the analysis of this case study. Entanglements are defined as 'intended or unintended, obvious or hidden, structured or chaotic interrelation(s) in space, knowledge or time'[3] and are, therefore, helpful conceptual tools to address the blind spots and blurred frontiers between media. As such, *Feu de Camp* is a potent example of entanglements between radio and comics. Nevertheless, the concept of media entanglements operates on a rather large scale, and two notions—intertextuality & intermediality[4]—are furthermore necessary for this analysis, as they operate on different levels, following the logic applied previously in this chapter. *Feu de Camp* was a radio show, however, it can only be understood fully through a cross-study of both radio and comics history. The show raises many questions, notably about the features resulting from these interactions between the two media. The general dynamic of entanglements found in *Feu de Camp* brought the world of *Pilote*—its stories, celebrities, and general spirit—into the one of radio. In addition, the radio programme also influenced, in return, and to a lesser extent, the magazine and comics in general. Therefore, this case study aims to uncover these interrelations, in examining a cooperative intermedial product, and to further develop an understanding of the connections between two popular media of the 1960s.

[3] Marie Cronqvist & Christoph Hilgert, "Entangled Media Histories. The Value of Transnational and Transmedial Approaches in Media Historiography", *Media History* (23:1, 2017), pp. 130–141.

[4] Jürgen E. Müller, "L'intermédialité, une nouvelle approche interdisciplinaire: perspectives théoriques et pratiques à l'exemple de la vision de la television", *Cinémas* (10:2–3, 2000), pp. 105–134.

1 Origins of *Feu de Camp*

Pilote was a weekly French magazine dedicated to *bandes dessinées* (the Franco-Belgian counterpart to comics) magazine, published from 1959 to 1989, which featured prominent comics series of the time (*i.e.* Achille Talon, Barbe-Rouge, and, probably the most famous one, Astérix) and published the works of key Franco-Belgian comics creators (*i.e.* Goscinny, Uderzo, and Gotlib). The magazine had a long-lasting impact on French popular comic culture, as many iconic series were born in the pages of *Pilote*. Its satirical social commentary, which had become the norm by the time the show was aired in 1969, can be considered to be its lasting legacy, however, as it inspired other prominent magazines such as *L'Echo des Savanes* and *Fluide Glacial*.[5] *Pilote*, which aimed at a teenage audience, was in competition with other similar magazines, like *Le Journal de Tintin* and *Spirou*, although they had a slightly younger readership. *Pilote* became monthly in June 1974, and published its last issue in October 1989. The birth of *Pilote* itself is significant here. The magazine was indeed born of the collaboration of comics creators, Albert Uderzo, René Goscinny, and Jean-Michel Charlier, in addition to Radio Luxembourg, a French-speaking commercial station, and some of its journalists. Interestingly, these ties operated on multiple fronts. The station partly financed the magazine, whose editor intervened on air on regular occasions to promote the upcoming first issue of *Pilote*.[6] The collaboration also included crossovers. The first was a comic adaptation of a radio soap opera, *Ca va bouillir*, by Zappy Max,[7] one of the most famous hosts of Radio Luxembourg in the Early Sixties. The show, in which Zappy Max was the main character, incorporated adverts for its sponsor, a detergent company, in the story. However, the comic adaptation, by

[5] Both magazines mentioned here were formed by *Pilote* creators, including Gotlib (co-host of this show), who wished to make their content more targeted to adults. This also went hand in hand with contemporary satirical magazines *Charlie Hebdo* and *Hara-Kiri*, whose creators would often also work for *Pilote*.

[6] Patrick Gaumer, *Les Années Pilote, 1959–1989* (Paris: Dargaud, 1996), p. 63.

[7] Max Doucet (1921–2019), better known as Zappy Max, was a French radio host. He joined Radio Luxembourg in 1947, where he presented many popular shows, most of them games, such as *Quitte ou Double*, *Le Crochet radiophonique*, and *Radio Circus*. When Radio Luxembourg became RTL in 1966 and went through many modernising changes, he was let go. He later joined another commercial station, Radio Monte-Carlo.

Maurice Tillieux[8] in the first issues of *Pilote* in 1959,[9] did not do so, putting some distance with the commercial aspect of the original radio show. It appears this 'medial transposition', to re-use a subcategory of intermediality suggested by Rajewski,[10] did not last long, however, this collaboration shows how strong the ties between commercial radio and *Pilote* were, already from its very beginning. The *Emission Pilote*, furthermore adapted various series, such as *Astérix, Tanguy et Laverdure,* and *The Demon of the Caribbean* to radio. *Pilote* was bought by Dargaud in 1960,[11] and the ties with Radio Luxembourg were gradually phased out. However, the magazine's ties with radio in general would endure, as *Pilote* was advertised on Radio Luxembourg's direct competitor, Europe n°1, furthering intermedial references. More precisely, it was advertised on the commercial station's famous programme *Salut les Copains* in 1966.[12] Links to Europe n°1 were to develop further in 1969, with the production of *Feu de Camp* as a common product, and the focus of this article.

Feu de Camp was hosted by René Goscinny (*Pilote* Editor-in-Chief), Gébé, Fred, and Gotlib, four key figures of the comics world at the time. Françoise Durgis, the only female voice of the programme, seconded the team, and was in charge of reading adverts, and acting as a representative of the station. A retrospective interview with Gotlib sheds light on his perspective on *Feu de Camp*. According to him, through transcribing the spirit of *Pilote* into radio, it gave the artists the opportunity to work together, and collaborate further, as they were usually working on their own with *Pilote*. Indeed, the sign-off catchphrase of the show stated that the entire team of *Pilote* was behind the production, as a collaborative group effort. In reality, only four of them were involved, although they invited other artists as guests. Furthermore, in the aforementioned interview, Gotlib insisted on how much they enjoyed themselves, and described the overall chaos and 'brouhaha' that came out

[8] Maurice Tillieux (1921–1978) was a Belgian *bande dessinée* artist, known for creating popular series like Gil Jourdan, and his collaboration with the Spirou magazine.

[9] Gaumer, *Les Années Pilote*, p. 66.

[10] Irina Rajewski, *Intermediales Erzählen in der italienischen Literatur der Postmoderne von den giovani scrittori der 80er zum pulp der 90er Jahre* (Tübingen: Gunter Narr, 2001).

[11] Sylvain Lesage, *Publier la bande dessinée: les éditeurs franco-belges et l'album, 1950–1990* (Villeurbanne: Presses de l'Enssib, 2018), p. 178.

[12] Salut les Copains, Europe n°1, 1966, 00:31:00, Archives of Europe 1, Paris.

of the programme. This might explain why, despite how successful they thought they were, Lucien Morisse, general manager of Europe n°1 at the time, called them after 13 episodes to say it was over as the show 'did not get the expected results'.[13]

Feu de camp was a rather atypical radio programme. It did not follow the usual adaptation of comics into radio, *i.e.* an audio dramatisation of a character's adventures. Nor was it following the habit of commercial radio stations to invite celebrities to host one of their shows.[14] Instead, the station gave *carte blanche* to a handful of prominent comic writers and artists to create their own new show, led by René Goscinny. The four men tried to recreate, on air, the spirit of their magazine and its humour. Due to this unusual, if not unique, mix of comics and radio, *Feu de Camp* can be described as a rather experimental product, questioning the links between the two media through a new lens. From the two surviving recordings of this programme, *Feu de Camp* revolved around a succession of short humoristic sketches, written and played by the creators. The show was also interspersed with songs (with lyrics echoing the themes of the sketches) and adverts (mainly for *Pilote* magazine), read by Françoise Durgis, the announcer. It appears *Feu de Camp* also included regular interviews with other artists from *Pilote*, who took part in the sketches. The hosts often incited listeners to buy issues of the magazine, implicating a secondary intermedial commercial goal behind this collaboration. The overall programme was meant as humoristic, as even the interviews were an opportunity for jokes, despite their apparent seriousness. Though the sketches seemed prepared, and potentially pre-recorded, the rest of the programme sounded like a live broadcast. It was also what could only be described as a chaotic kerfuffle of voices, jokes, and interruptions, close to the 'brouhaha' described by Gotlib.

[13] Interview with Gotlib about René Goscinny, extract from *René Goscinny, profession humoriste* (1998), https://vimeo.com/194408112 (last consulted 07/March/2020).

[14] For examples see, French *yéyé*-singer Claude François who presented Europe n°1's popular *Salut les Copains* in 1965, or *Le Journal inattendu* on RTL, in which various public figures took over the role of editor of Saturday's midday news. See Salut les Copains (présenté par Claude François), Europe n°1, April 1965, 00:39:21, Archives of Europe 1, Paris, & Denis Maréchal, *RTL. Histoire d'une radio populaire* (Paris: Nouveau Monde, 2010), p. 362.

2 Key Features

Considering the unique nature of a radio programme at the crossroads of two media, it is worth looking into the resulting key features that were born out of this intermedial status. In the first instance, the amateurism of the hosts appeared quite clearly, as they demonstrated they were not familiar with radio norms; or chose not to follow them. For instance, the hosts had a tendency to talk over one another, to interrupt the person reading the adverts, and try to make her laugh on air. One example of this amateurism is particularly striking when they receive Jean Tabary as a guest, and forget to properly introduce him for listeners.[15] Instead a cacophony of laughs, jokes, and overacted surprised gasps is heard, and Goscinny has to correct the situation. This shows how the hosts forget a rather basic rule of radio: listeners cannot see what is happening in the studio. Tabary also sounds uncomfortable in front of a microphone; his voice is barely audible, he stutters, pauses, and loudly turns pages, for instance. In addition, he is often interrupted by others, notably Goscinny who tells him, on-air, to articulate. Amateurism runs through the show, especially because the hosts seem to forget they are on air, and that listeners are tuning in. It might be explained by the fact the hosts were comic creators who attempted to transfer the spirit of their magazine into radio, without the experience of its rules and norms. The Tabary interview reveals another characteristic of the programme; its playfulness. Indeed, throughout the recordings, it appears the hosts often tease one another and engage in self-mockery. In the case of Tabary, he is asked to read a script to promote the magazine, but the *Pilote* team had thrown in tongue-twisters, as well as overly complicated French grammar, to make him slip up on air. The result is a rather hectic discussion, with voices overlapping one another, and regular laughter. Playfulness, and humour in general, are therefore a key feature of *Feu de Camp* and of *Pilote*, running through the magazine in its strips and editorial content. As such, this can be deemed as the most striking aspect of the spirit that they wanted to bring with them from the magazine.

The combination of amateurism and playfulness described above could be an explanation of a third feature of *Feu de Camp*: its chaotic atmosphere. Instead of a structured radio programme, *Feu de Camp* comes

[15] Le Feu de camp du dimanche matin, 2nd November 1969, Europe n°1, 00:46:39, Europe 1 archives, Paris.

across as friends gathering for a bawdy chat. As mentioned, they forget, or give the impression they forget, the existence of listeners. The one semblance of control comes from Goscinny, who acts as a paternalistic figure in the programme, bringing order to discussions, and regularly joking he will call some of them into his office. His leading position in the show is interesting, as it is inherited from *Pilote*, where he was Editor-in-Chief, despite the four creators being billed as co-hosts of *Feu de Camp*. This echoes a power struggle that occurred the previous year, when *Pilote* took on a greater anti-establishment stance after the events of May 1968,[16] when creators revolted against Goscinny and obtained more creative liberty, although were ultimately still under his authority.[17] Despite this chaos not being particularly radiogenic, it was such a defining trait of the programme that it was used as a selling point when advertised in *Pilote*.

Promotion for the show and its chaotic nature can be found on several occasions. Indeed, illustrated examples of advertisements show the group around an actual campfire that sets the Europe n°1 studio on fire.[18] Another photographic advert plays on the unprofessionalism by using a *roman photo*.[19] It shows the creators, in the studio, annoying the station's announcer by getting overwhelmingly tangled in the cords around her, and spilling onion soup. The advert playfully uses the onion soup as the point of interest, telling readers: 'if you want to hear [...] the *Pilote* team eating onion soup on the radio, listen every Sunday [...] *Le Feu de Camp du dimanche matin*'.[20] This advert reflects the show itself, with the *Pilote* team often annoying the announcer while she read the adverts, as well as their uncouth manners of snorts and hacking coughs on air. Therefore, chaos was not only accidentally born out of their amateurism and playfulness, it was something they embraced, and used in the magazine to

[16] Wendy Michallat, *French Cartoon Art in the 1960s and 70s: Pilote hebdomadaire and the Teenager Bande Dessinée* (Leuven: Leuven University Press, 2018), p. 169.

[17] See Eric Aeschimann & Nicoby, *La Révolution Pilote: 1968–72* (Paris: Dargaud, 2015).

[18] *Pilote*, n°522, 6th November 1969; and *Pilote*, n°530, 1st January 1970.

[19] Photos placed into order to make comic panels which then have speech balloons added, and therefore form a story, a storytelling technique often used in youth magazines at the time.

[20] *Pilote*, n°529, 25th December 1969.

encourage people to tune in. Nevertheless, based on Gotlib's interview, this taste for chaos might not have been shared by Europe n°1.[21]

Commercialism, as suggested by the aforementioned adverts, is another shared feature found in *Feu de Camp* to be discussed here. Indeed, both Europe n°1 and *Pilote* were commercial entities run as businesses. Adverts are therefore the most tangible form to be considered for this article, and they have to be deconstructed into two categories. First, there were 'general' adverts, common to Europe n°1, read by Françoise Durgis, for various products, such as shampoos. Second, *Pilote* and *Feu de Camp* regularly advertised one another. In the magazine, adverts could take the form of the *roman photo* mentioned previously, but were usually illustrations by *Pilote* artists, promoting *Feu de Camp*. While on the radio show, the announcer read adverts such as: '*Pilote*. A good laugh is worth more than a good steak! So, devour *Pilote*! *Pilote* magazine—new look, large format, bursting with humour and hilarious jokes!'.[22] Moreover, as previously stated, the hosts and their guests regularly mentioned the magazine and encouraged listeners to buy it. Looking at adverts offers insights into the shared commercial nature of both media, as they similarly integrated, despite their differences, adverts into the fabric of their outputs. It also clearly reveals just how entangled *Feu de Camp* and *Pilote* were, as they worked hand in hand to promote one another through a variety of intermedial references.

3 'We Make Radio Because It Gives Our Eyes a Rest'

It is important to note that the creators themselves were aware of the cross-medial status of the programme. Indeed, they advertised this unique aspect by playing on the tensions it provoked between radio and the magazine. For instance, the tagline of *Feu du Camp* itself, '*Nous, on fait de la radio parce que ça nous repose les yeux*' (We make radio because it gives our eyes a rest), infers the lack of direct images in radio, especially in opposition to comics. In typical *Pilote* humouristic style, the programme on several occasions claimed to be 'the most visual of radio shows', both

[21] This sheds light on a possible reason why the show only lasted 13 episodes. Furthermore, a hypothesis raised by Europe 1's archivist to explain the chaos was the possibility they were not sober while recording, something which could not be ruled out.

[22] Le Feu de camp du dimanche matin, 23rd November 1969, Europe n°1, 01:22:11.

on air and in printed adverts. This is once again a play on the essence of radio, and how they were trying to subvert its norms. Comics (and by extension, *Pilote*) is a highly visual medium, and, by bringing its spirit onto radio, *Feu de Camp* asserts the idea it could overcome this lack of direct visuals. In addition, they also acknowledge the lack of sound within comics.[23] The first advert for the show that appeared in *Pilote*, drawn by Uderzo, reflects the 'silence of comics'. The illustration shows Fred, Goscinny, Gébé, and Gotlib seated at a banquet table around a campfire (as an intermedial reference to the numerous banquet scenes in the *Astérix* series), with Goscinny speaking animatedly into his microphone. In the foreground sits Uderzo, the *Astérix* artist himself, tied and gagged with a bandana, as regularly happened to the village bard in the series. Incidentally, Uderzo's signature is also drawn with a gag around it. He was an artist, in contrast to the rest of the creators on the show who were also writers. This could be a link to the lack of sound in comics' imagery, with the artist becoming voiceless without his main means of creative expression.

Momentarily setting aside the notion of whether or not the show managed to overcome the limitations of its media of origin, it is nevertheless interesting to comment on the fact it was sold as a product capable of bridging the two media. This points to the idea that the people behind *Feu de Camp* thought of this intermedial sonic experiment as something exciting and marketable. A closer look at the surviving recordings, however, shows the limits of this experiment. In addition to the chaos and amateurism mentioned previously, many jokes found in the sketches are inspired by comics gag setups and rely on visual twists. As a result, they are usually described by one of the hosts, unfortunately making the jokes particularly lengthy as many details are required to overcome the lack of images.[24] One demonstration is found in a sketch in which Fred engages in fire-breathing, to prove the 'visual power of the show'. However, this allegedly goes poorly and fire burns off Fred's hair and moustache, which makes him unrecognisable. As a consequence, the other hosts have to

[23] Silence of comics is a contested concept, with the implication that there is no physical sound. However, it is acknowledged that readers will often 'hear' voices and sound effects. See Mark Grimshaw-Aagaard, Mad Walther-Hansen & Martin Knakkergaard (eds.), *The Oxford Handbook of Sound and Imagination Volume 1* (Oxford: Oxford University Press, 2019), p. 514.

[24] Le Feu de camp du dimanche matin, 23rd November 1969, Europe n°1, 01:22:11.

over-exaggerate to explain his physical changes for listeners to understand. Arguably, this would have worked more effectively in a visual format, easily shown in a comic through one image. Examples of radio adaptations of *bande dessinée* stories (*i.e.* Astérix and Tintin) show that crossovers between the two media are possible. Nevertheless, *Feu de Camp* shows that adaptation from a highly visual primary cultural material into audio is not necessarily always a success. As such, despite the initial claim of *Feu de Camp* being 'the most visual radio show', it does not consistently deliver on this promise.

Perhaps the most significant successful transmedial exchange is that of *Bougret & Charolles*. A short radio drama which was made for *Feu de Camp*, the sketch later became a comic series of its own, thereby making it a prime example of intermediality. It furthermore deferred from the usual practice of adapting comics to radio. *Bougret & Charolles* was created by Gotlib, and centred around a detective (Bougret, played by Gébé) and his hapless sidekick (Charolles, played by Gotlib), who solved crimes using ridiculous logic, while secondary characters were played by Fred and Goscinny. The investigations of Bougret and Charolles were quite radiogenic and fit in well with the radio drama format. For instance, they made good use of the voice of a narrator, various sounds of typewriters to illustrate a police station, music echoing the film noir tradition, etc. The stories are riddled with intertextual references to a wider literary corpus of police mysteries, such as the Sherlock Holmes series. For instance, Charolles holds the role of the narrator, as did John Watson, and the catchphrase 'Elementary, my dear Charolles' is used. Furthermore, it references Jules Maigret, the police commissioner created by Georges Simenon. Bougret, whose name echoes Maigret's, is heard smoking the pipe on a few occasions, just like the fictional policeman. It is interesting to note that a police mystery parody already existed on Europe n°1: *Signé Furax*, which was a popular programme in the late Fifties. *Bougret & Charolles* was conceived for radio, rather than adapted, and played on existing tropes. The episode also has a self-contained story, or investigation, rather than stretching it out over several episodes.

What is particularly fascinating is that Gotlib later published *Bougret & Charolles* in *Pilote* as a comic strip, as part of the *Rubrique-à-Brac* current affairs segment. He kept the appearances of his fellow creators and radio hosts as what the characters 'looked' like.[25] The characters were thus caricatures of the artists. As such, *Bougret & Charolles*, in both radio and *bande dessinée* forms, constitute an example of hybridisation, illustrated for instance in the artists, who gave either their voice or their face to the characters. The comic strip had a lasting cultural impact on *Pilote*, even appearing as the magazine's cover image.[26] It also influenced the broader French comics market when published in albums. The purpose of *Feu de Camp* was to bring the spirit of the magazine on air; however, *Bougret & Charolles* went in the opposite direction and had more enduring success than the programme. One hypothesis can be formed. Content-wise, and based on surviving recordings, it was the most radiogenic element of the show, which might explain why the stories survived and carried on into *Pilote*. It is the most interesting case of 'entanglements' within the show, as a radio-born adventure found its place in the magazine, creating a loop feeding back into the comics realm.

This chapter has examined the outcomes of *Le Feu de Camp du dimanche matin*, a specific transmedial experiment which aimed to bridge the gap between the media of comics and radio. Throughout the research undertaken here, *Feu de Camp*, alongside several issues of *Pilote*, has been the main object of study, as they connected the various media worlds. A series of key features, which resulted from the transmedial experiment, arose from the analysis. While some similarities are shared between the magazine and the radio station, such as the commercial nature and the omnipresence of adverts, others originate from the medial transposition from comics to radio. In this regard, the characteristic playfulness of the *Pilote* team transferred with ease onto the airwaves of the commercial station. Nevertheless, the transposition also shaped the programme by a form of amateurism and some chaos, which can be noted as striking features of *Feu de Camp*. It would seem, however, that some elements of the famous satirical visual style *Pilote* was known for simply did not travel well to radio. The possibility of countering the lack of direct imagery in radio and the lack of sound in comics could not necessarily be achieved by a mere combination of the media. There were, however, some elements that worked well, including the *Bougret & Charolles* series. This series

[25] In addition, the culprit and the innocent man, in the comics, were named after the artists Fred and Gébé's real names.

[26] *Pilote*, n°649, 13th April 1972.

was a particularly noteworthy aspect, as it exists as a rare demonstration of the possibility of radio to transfer to comics, rather than the other way around, creating a fascinating example of media combination. Overall, *Feu de Camp* was also a product rooted in its time. The late 1960s was a period of great popularity for each of the media, and it is therefore interesting that they would combine forces for a collaborative and commercial project. Europe n°1 was in a strong position for listenership in France, meanwhile *Pilote* was selling a large number of copies, meaning that their intermedial product represents a fascinating gateway into the historiographical blind spots between the two media.

References

Aeschimann, Eric, and Nicoby. *La Révolution Pilote: 1968–72*. Paris: Dargaud, 2015.

Cronqvist, Marie and Christoph Hilgert. "Entangled Media Histories. The Value of Transnational and Transmedial Approaches in Media Historiography". *Media History* 23:1 (2017): 130–141.

Gaumer, Patrick. *Les Années Pilote, 1959-1989*. Paris: Dargaud, 1996.

Grimshaw-Aagaard, Mark, Mad Walther-Hansen and Martin Knakkergaard, eds. *The Oxford Handbook of Sound and Imagination Volume 1*. Oxford: Oxford University Press, 2019.

Legay, Richard and Jessica Burton. "From the Comic Strips to the Airwaves. The short-lived experiment of Le Feu de Camp du Dimanche Matin on Europe n°1", *Media History* 29:2 (2023): 226–239.

Lesage, Sylvain. *Publier la bande dessinée: les éditeurs franco-belges et l'album, 1950-1990*. Villeurbanne: Presses de l'Enssib, 2018.

Michallat, Wendy. *French Cartoon Art in the 1960s and 70s: Pilote hebdomadaire and the Teenager Bande Dessinée*. Leuven: Leuven University Press, 2018.

Müller, Jürgen E. "L'intermédialité, une nouvelle approche interdisciplinaire: perspectives théoriques et pratiques à l'exemple de la vision de la télévision". *Cinémas* 10:2–3 (2000): 105–134.

Rajewski, Irina. *Intermediales Erzählen in der italienischen Literatur der Postmoderne von den giovani scrittori der 80er zum pulp der 90er Jahre*. Tübingen: Gunter Narr, 2001.

CHAPTER 11

The *Dispositif* of Commercial Radio Stations

While conducting research on the history of commercial radio in the Long Sixties, one can only be impressed by the sheer amount of contemporary nostalgia in regards to Radio Luxembourg and Europe n°1, especially a show like *Salut les Copains*. Whether it is grey literature, social media pages, or the short-lived experiment of 'United DJs—The Station of the Stars',[1] there are many examples of nostalgic practices in regards to commercial radio stations. This current nostalgia and presence of commercial radio stations in memory can be explained, to some extent, by the impact the stations had on their listeners and their readers throughout the Long Sixties. If the stations had such an impact, it is because they consciously operated on various levels, which were studied in the previous chapters. The analyses conducted previously in this book have all pointed to the existence of a wider infrastructure, a larger apparatus—or system—at stake behind it all.[2] In other words: a *dispositif*, or an ensemble

[1] United DJs was an English-speaking webradio launched by former radio hosts of the English service of Radio Luxembourg in order to recreate the perceived good old days of radio. It was online between 2018 and 2021.

[2] The knowledge—or, at least, the intuition—of the existence of a wider *dispositif* operating with radio was already present in the 1930s, as illustrated in a 1933 study quoted by Taylor, Katz and Grajeda: 'When the average listener goes to his radio, turns it on and idly spins the knob that brings him first one station and then another, he has little or no conception of the work, the energy, the care and the manifold talents going into

R. Legay, *The Cultural Histories of Radio Luxembourg and Europe n°1*,
https://doi.org/10.1007/978-3-031-46250-4_11

able to bring together the aforementioned dimensions, which explains how commercial radio stations could have had such an impact on their listener-readers.[3]

In order to understand the *dispositif* of commercial radio stations, this chapter details seven of its constitutive elements. All of them were present—one way or another—in previous chapters, and are the ideal way to analyse an ensemble otherwise hard to grasp in its entirety. Therefore, they act as seven 'nods' where the various dimensions of the concept interact. Following the works of other scholars on media history, the *dispositif* was split in three dimensions: materiality, content, and perception.[4] However, due to the nature of Radio Luxembourg and Europe n°1, a fourth dimension should be considered here: the commercial dimension, an echo of the underlying 'strategic function'.[5] As commercial broadcasters, the primary goal of Radio Luxembourg and Europe n°1 was—and still is, for they are run as businesses—'the relentless drive to cut costs and increase profits'.[6] The four dimensions of the analysed below are

the making of the various programs he so lightly considers and dismisses by a move of the dial', Gustav Klemm, "Putting a Program on the Air", *Etude*, March 1933, in T. D. Taylor, M. Katz & T. Grajeda (eds.), *Music, Sound, and Technology in America: A Documentary History of Early Phonograph, Cinema, and Radio* (Durham: Duke University Press, 2012), p. 344.

[3] In Giorgio Agamben's vision, the dispositif is 'anything which has, one way or another, the capacity to capture, guide, determine, intercept, shape, control, and ensure the gestures, behaviors, opinions and discourses of living beings' Giorgio Agamben, *Qu'est-ce qu'un dispositif ?* (Paris: Payot, 2007), p. 30.

[4] Markus Stauff details how the tool of the *dispositif* has been used, pragmatically, to study 'the materiality of a medium, the content it mediates and the perception or symbolic meanings it conveys'. Markus Stauff, "Materiality, Practices, Problematization", in S. Aasman, A. Fickers and J. Wachelder (eds.), *Materializing Memories: Dispositifs, Generations, Amateurs* (New York: Bloomsbury, 2018), pp. 67–83. An approach found, for instance, in Van der Heijden's work, whose take on the concept leads him to define three dimensions: material, textual, and perceptual. Tim van der Heijden, *Hybrid Histories: Technologies of Memory and the Cultural Dynamics of Home Movies, 1895–2005* (Maastricht: Universitaire Pers Maastricht, 2018), p. 45.

[5] 'The Confession of the Flesh' (1977), interview with Michel Foucault, in Michel Foucault and Colin Gordon (ed.), *Power/Knowledge: Selected Interviews and Other Writings* (Harlow: Longman, 1980), pp. 194–228.

[6] Hugh Chignell, *Key Concepts in Radio Studies* (London: Sage, 2009), p. 114.

not immutable, and were defined as 'mutually inclusive',[7] reinforcing the interconnections between the seven elements.

1 The Transmitter Sites

The transmitter sites, core elements of the broadcasting apparatus, are the first logical step of the analysis of the *dispositif*. As mentioned by Steven Connor: 'the apparatus of radio has slipped out of the picture, as radio has got the reputation of being an immaterial art',[8] showing the importance of bringing this apparatus back into the picture, a process which starts with the transmitter sites. This first section largely builds on the analysis conducted in the first three chapters to present transmitter sites as entry points into the material and symbolic dimensions. As we saw, the material anchoring of commercial stations—particularly through their transmitters—is fundamental in establishing the appropriation of a transnational broadcasting space. Indeed, the transmitters' interrelations with the studios and offices were all transnational flows allowing the stations to broadcast. They were also decisive in defining their nature as commercial broadcasters—or *radios périphériques*—from a legal perspective.[9] The anchoring of transmitter sites in specific places—the Saarland and Luxembourg—is the first large-scale structural element of the *dispositif*, the one that enables broadcasting. This element is not only large-scale, it also represents a rather static point throughout the period. Even if the power of the transmitters evolved, and the quality and comfort of listening depended on various elements, the existence of transnational interrelations between studios, transmitters sites and offices were always a given in the long 1960s. It also points at a key difference between public and commercial stations: the need for Radio Luxembourg and Europe n°1 to have buildings—offices, studios, transmitters—in multiple

[7] Andreas Fickers, "'Neither Good, nor bad; nor neutral': The Historical *Dispositif* of Communication Technologies", in M. Schreiber and C. Zimmermann (eds.), *Journalism and Technological Change: Historical Perspectives, Contemporary Trends* (Frankfurt: Campus, 2014), pp. 30–52.

[8] Steven Connor 'I Switch Off', in D. R. Cohen, M. Coyle and J. Lewty (eds.), *Broadcasting Modernism* (Gainesville: University Press of Florida, 2009), pp. 274–293.

[9] Indeed, as mentioned previously, it is because of their transnational nature, being established as *radios périphériques*, that Radio Luxembourg and Europe n°1 could circumvent State monopolies of neighbouring countries.

countries, something public broadcasters were not concerned with. Both the maps of Europe n°1 and the minutes of meetings of Radio Luxembourg studied in the first chapters show how crucial transmitters were for commercial radio stations, making them a key element. They were also central in placing the stations in the wider European broadcasting space, as illustrated by the tensions around Europe n°1's frequencies.[10]

When explaining the *dispositif*, Foucault mentions architecture.[11] If connections with the carceral system he studied in *Surveiller et Punir*[12] (Discipline and Punish) appear rather clearly, the question of architecture could seem, at first glance, slightly incongruous when raised in the context of commercial radio. Nevertheless, architecture has always represented, for broadcasting, a crucial stake. In the case of transmitter sites, Europe n°1's in Saarland, on the Felsberg plateau, with its modern design, made of glass and concrete, shaped like a mussel, represents a powerful example. According to Maryam El Moumni, who studied the building as a Telecommunications Heritage Site, the significance of the Saarland transmitter can be read through multiple lenses.[13]

The building which hosted the transmitter started broadcasting in 1955, up until 2015, when it was subsequently sold in order to be preserved to the local municipality of Überherrn, with subvention from the German Federal Ministry of Interior.[14] The building is of particular architectural interest, especially the prestressed concrete shell-roof,

[10] Indeed, since its first broadcast and until 1975, Europe n°1 used non-legitimate frequencies until an official one was attributed at the Geneva broadcasting conference. Although the situation was not completely uncommon, it still provoked many international protests and pressures. See Andreas Fickers, "Die Anfänge des kommerziellen Rundfunks im Saarland", in C. Zimmermann, R. Hudemann and M. Kuderna (eds.), *Medienlandschaft Saar von 1945 bis in die Gegenwart. Band 1: Medien zwischen Demokratisierung und Kontrolle (1945–1955)* (München: R. Oldenbourg, 2010), pp. 241–308.

[11] More precisely, Foucault actually mentions 'architectural forms'. "The Confession of the Flesh", in Foucault and Gordon, *Power/Knowledge*, pp. 194–228.

[12] Michel Foucault, *Surveiller et Punir: la naissance de la prison* (Paris: Gallimard, 1975).

[13] Maryam El Moumni, "The Cultural Significance of Telecommunication Heritage Sites: The Case of Europe 1 Transmitter Building in Berus, Saarland", Master diss. (Brandenburg University of Technology Cottbus-Senftenberg & Deakin University, 2017).

[14] Johannes Bodwing, "Was macht Überherrn mit diesem Juwel?", *Saarbrücker Zeitung*, 3rd August 2016, [https://www.saarbruecker-zeitung.de/Saarland/saarlouis/ueberherrn-mit-diesem-juwel_aid-56919] (last accessed 14 March 2023).

which gives the building a 'distinguished place in the history of architecture and engineering of the twentieth century'.[15] Interestingly, according to Ericson and Riegert, the advent of radio and television broadcasting, and the idea of materiality being dissolved 'into the air', had an impact on architecture, by implementing themes such as transparency and lightness.[16] The Saarland transmitter, with its tall glass windows and wide main hall, does reflect this. Part of the reasoning behind such a vast main hall was to allow 'the accommodation of successive transmission systems in different blocs',[17] inferring the question of a technical upgrade being integrated in the building from its early days. As the study on the maps from the transmitter site revealed, the strength of the signal emanating from the transmitter played a key role in the appropriation of the transnational broadcasting space. The area covered and the quality of the signal were both important for the station, factors often measured by the technical staff. Their changes over time, as shown by the maps, in addition to the building's upgrade, point to the evolution of the *dispositif*, with the aim of expanding their reach into the transnational space from the place of the transmitter site.

Carrying on with the idea that the various dimensions of the *dispositif* are inclusive, it is also relevant to include the idea developed by French geographer Michel Lussault of referring to all sorts of authorities (understood in a large sense) as 'agencies',[18] which makes sense when looking into commercial radio stations. While the historiography on Europe n°1 is not well-developed, the history of Radio Luxembourg, especially its French service, has been at the centre of attention of more historical works. Generally, the stations have been studied as such agencies, and the literature has shown how the stations' organisation evolved from their early days and throughout the twentieth century. What is striking

[15] El Moumni, "The Cultural Significance of Telecommunication Heritage Sites", p. 62.

[16] Staffan Ericson and Kristina Riegert (eds.), *Media Houses: Architecture, Media, and the Production of Centrality* (New York: Peter Lang, 2010), pp. 4–12.

[17] El Moumni, "The Cultural Significance of Telecommunication Heritage Sites", p. 81.

[18] Michel Lussault, "Renouveler le dialogue", *Espace Temps* (68:70, 1998), pp. 31–44. For more details about the use of the concept of *dispositif* by geographers in France, see Marc Dumont, "Aux origines d'une géopolitique de l'action spatiale: Michel Foucault dans les géographies françaises", *L'Espace Politique* (12:3, 2010), [https://doi.org/10.4000/espacepolitique.1744] (last consulted 5 August 2020).

in this historiography is the richness of the connections with a multitude of other agencies (*e.g.* advertisers, companies) and institutions (*e.g.* various European States, the European Broadcasting Union), born out of the commercial and transnational nature of the stations. This is quite revealing of the intense network of exchanges and actors in which one finds commercial radio stations entangled. This intense network is also the main producer of a large body of 'laws and regulatory decisions'—to use Foucault's phrasing—or the 'binding framework of the dispositif'[19]—to use Larroche's—within the internal organisation of the stations,[20] and in a wider transnational context, reflecting the nature of the agencies. It also shows that the *dispositif* of commercial radio stations is itself linked to a wider European broadcasting space. Moreover, this 'binding framework' had an impact on the establishment of both of the stations, whether it is their infrastructure, transmitter sites, offices, studios, and even the content they produced.

2 The Studios

While the transmitter sites play a crucial role in the *dispositif*, their role only makes sense in relation with the studios, where the broadcast programmes are produced.[21] Radio programmes made in studios were broadcast live or pre-recorded. In this regard, Radio Luxembourg played a key role in introducing recorded programmes on the radio in the 1930s, most notably with its famous show *The League of Ovaltineys*, made in London and sent to Luxembourg.[22] Studios exist, among other reasons,

[19] Valérie Larroche, *The Dispositif: A Concept for Information and Communication Sciences* (London: ISTE, 2019), p. 59.

[20] This echoes the organisational and hierarchical structure of the stations as commercial enterprises, as well as internal rules. In the case of what is studied here, two examples are of particular interest. The pro-European stance of RTL for instance, which was a general guideline for reporters and journalists that shaped the image and discourse of the station. The second example is the controversial and provocative figure of Jean Yanne. Indeed the media personality and comedian worked on various stations and was let go for his tone.

[21] Not all radio programmes were recorded in studios, some could be made elsewhere, in particular in reporters' work, as well as some game shows (*Radio Circus* and *Balzac 10-10*, mentioned previously, for instance). However, in this context, only those who were made in studios are considered.

[22] Sean Street, *Crossing the Ether: British Public Service Radio and Commercial Competition, 1922–1945* (Eastleigh: John Libbey, 2006), p. 112.

to provide an environment with ideal acoustics[23] for radio production, which is determined by the reflective properties of surfaces and the position and types of microphones.[24] Moreover, studios help reinforce the sense of intimacy between hosts and listeners: 'if the sounds are produced in a studio and all resonance is deadened then these sounds seem to occupy the same space as that of the listener, replicating the acoustic qualities of most people's homes, where typically sounds are deadened by carpets, wallpaper, curtains, and furniture'.[25] In the studio, presenters and especially disc-jockeys are at the centre of the programme and they face both the 'corporate boss' and the listener,[26] a dynamic materiality represented by the typical layout of studios, where disc-jockeys present the programme from a central desk.[27] Therefore, studios represent a particularly important cog in the wider *dispositif* of commercial radio stations, however, their role was not only material, but also highly symbolic.

According to Chignell, studios are the place from where radio hosts rule over the 'radio worlds' they create on air,[28] suggesting another bridge between the material, content, and symbolic dimensions. The presence of the studio on air is paradoxically heard, most of the time, through its absence due to the aforementioned 'deadened sounds'. Nevertheless, as Chapter 6 has shown, the sound of commercial radio stations also included sonic elements which pointed at the material experience of listening. In numerous programmes, listeners could hear a variety of

[23] Acoustics are understood here as 'the quality and nature of sound in a particular physical environment', as defined by Chignell, *Key Concepts in Radio Studies*, pp. 7–9.

[24] Guy Starkey, *Radio in Context* (Basingstoke: Palgrave Macmillan, 2004), p. 11. For a detailed account of the history of technology in regards to studios and recording material, see J. Harman, "Recording and Studio Equipment", in Christopher Sterling (ed.), *The Museum of Broadcast Communications: Encyclopedia of Radio*, (New York: Fitzroy Dearborn, 2004), pp. 1187–1193.

[25] Martin Shingler and Cindy Wieringa, *On Air: Methods and Meanings of Radio* (London: Arnold, 1998), p. 56.

[26] Chignell, *Key Concepts in Radio Studies*, pp. 17–21.

[27] Starkey, *Radio in Context*, p. 63.

[28] The 'radio world' is a concept based on the discursive space/imaginary world created and ruled by radio hosts, which was rephrased by Chignell. A well-known example is found in analysis of the *Tony Blackburn Show* in the 1980s by Brand and Scannell. Chignell, *Key Concepts in Radio Studies*, pp. 95–96, and Graham Brand, and Paddy Scannell, "Talk, identity and performance: The Tony Blackburn Show", in Paddy Scannell (ed.), *Broadcast Talk* (London: Sage, 1991), pp. 201–226.

sounds which reminded them of the context in which the programme was made, such as sounds of chairs or papers being moved, usually reinforcing the feeling of the liveness of the programmes. Therefore, studios, as places of production, are key nods to the *dispositif*, in which radio hosts have a particularly powerful role, at the crossroad of materiality, content, and symbolism of commercial radio.

Studios can also be understood in a wider sense, by metonymy, as the buildings hosting the places of production described above. In this regard, the importance of architecture mentioned by Foucault plays a central role in highlighting the strong links between the material and symbolic dimensions of the studios, which are often known to the public. In this regard, the case of State broadcasters might shed some light on the phenomenon. Architecture in relation to broadcasting was of key importance for the French and British States for instance. This appears rather clearly with the examples of Broadcasting House[29] in London and of the *Maison de la Radio*[30] in Paris, which are particularly revealing of the importance of buildings for State broadcasters. This importance was incarnated through multiple forms. These buildings were not only practical—hosting studios and offices—they also had a centralising function, considering they were holding prominent institutions of both British and French States. Moreover, such buildings are particularly integrated into the symbolic dimension of the *dispositif*, as they held a symbolic status as well, for they were the fronts of their respective institutions, the British Broadcasting Corporation and its French counterpart, the (Office de) Radiodiffusion Télévision Française.

[29] The Broadcasting House was built in 1932 to host the headquarters of the BBC, right in the centre of London. The building, designed by George Val Myer, was inspired by the Art Deco style and became architecturally iconic. It is also a listed building, meaning any changes to it need to be specifically approved, in order to protect its cultural value. For more details on the history of the building, see Mark Hines, *The Story of Broadcasting House: Home of the BBC* (London: Merrell, 2008).

[30] The *Maison de la Radio* (House of Radio in English) was designed by French architect and urbanist Henry Bernard after he won a competition launched by the French State. The building was inaugurated in December 1963 by French president Charles de Gaulle in a heavily mediatised ceremony (See Arrivée du Général de Gaulle pour l'inauguration, ORTF, 14th December 1963, 00:01:36, INA.fr [https://www.ina.fr/video/i00011617], last consulted 14 March 2023). The building was labelled in 2018 a *monument historique*, an official designation in France for national heritage sites, reinforcing its architectural and cultural value.

Regarding the offices of both commercial radio stations in Paris, the analysis of the radio programmes has shown how much the Parisian addresses of the stations were integrated into the content of the shows. Not only were the addresses said time and again, they were also turned into jingles, reinforcing the power of the message through non-discursive sound elements. Sonic icons were key features of the sound of commercial radio stations in the 1960s, and, therefore, the geographical locations of the aforementioned buildings became symbolic[31] through their integration into the sound of the stations. The example of the jingle '22, rue Bayard' in Rosko's show *Minimax* on RTL and is a perfect illustration of the point developed so far. The address refers to the historical headquarters of the French service of the station, from 1936 to 2018,[32] which became associated with the broadcaster. The building itself was subject to important renovation works in 1966, following the drastic changes and new orientation of the station that same year. Interestingly, the modernisation concerned the programmes, the staff, as well as the very building hosting the station, from then on known as RTL, showing the connections of various elements of the *dispositif* of the commercial broadcaster. Moreover, the renovation of the building on Rue Bayard also concerned the offices, notably the director's, which became a paragon of 1960s design, the studios—notably Rosko's, which became very modern and automated—and, finally, the facade of the building, the public face of the station in the streets of Paris.[33] Therefore, it is possible to see how one building—in other words, an expression of the architectural form mentioned by Foucault—can carry multiple significations and impact other dimensions of the *dispositif*. The '22, rue Bayard' was, all at once

[31] According to Larroche: 'the mediation dispositive is a space making it possible to link an operational and material dimension to a symbolic dimension', a statement that applies especially well on the aforementioned buildings, which operate on these two dimensions. Larroche, *The Dispositif*, p. 68.

[32] Sarah Lecoeuvre, "RTL: le siège historique du 22, rue Bayard, n'existe plus", LeFigaro.fr, 1st June 2019, [https://tvmag.lefigaro.fr/programme-tv/rtl-le-siege-historique-du-22-rue-bayard-n-existe-plus_abc4e4c0-8461-11e9-bc1d-9b7a6df1c27b/] (last consulted 14 March 2023).

[33] See Jean-Pierre Defrain and Jacques Boutelet, *RTL. 40 ans ensemble* (Paris: Calmann-Lévy, 2006), pp. 19–22. Moreover, the facade was moved to a museum on art and architecture—the Fondation Vasarely in Aix-en-Provence—when the building was demolished, reinforcing the cultural and artistic power and value of this part of the building.

(1) a cog in the enablement of a transnational broadcasting system, (2) a technical and institutional hub for radio staff, (3) a place of power, both internally as headquarters of the French service of Radio Luxembourg, and externally, as a public face in the French capital, (4) the material incarnation of the wave of changes that hit the station in 1966, (5) a sonic icon, a crucial piece of the sound of commercial radio stations, and, (6) it could be argued that the address, with all it signified, even acted as one of Pierre Nora's *lieux de mémoire*.[34]

Despite the focus on the Parisian examples, it does not mean that the Villa Louvigny, which hosted the studios of Radio Luxembourg in the heart of the Grand Duchy, did not have symbolic power too. The site was also particularly significant in the history of the station and a *lieu de mémoire* in Luxembourg.[35] Indeed, the Villa Louvigny—among other things—hosted the *Eurovision Song Contest* in 1962 and 1966, and famous concerts such as the Grateful Dead and the Beach Boys in 1972, events which contributed to the symbolic construction of the site in Luxembourg. In September 2018, the Villa Louvigny officially became a heritage site in Luxembourg, and, on 29 September 2019, the building hosted a series of three concerts as part of the national celebrations for the Heritage Days. Therefore, studios—understood as the recording place, and, by extension, the building where radio is made—constitute a key element and nod of the *dispositif* of commercial radio stations, as places of production on one hand, but also, on the other hand, as symbolic places.

[34] The concept developed by French historian Pierre Nora of *lieu de mémoire* is vast and has led to many additional writings, nevertheless, it could be explained in a few words, as a place (literally or symbolically) where memory 'crystallizes and secretes itself'. Therefore, this seems an adequate way to describe what occurred to the '22, rue Bayard', which occupies a particular place as a *lieu de mémoire* in popular memory in France. Interestingly, the address plays this role on both literal and symbolic levels. See Pierre Nora, 'Between Memory and History: "Les Lieux de Mémoire"', *Représentations* (26, 1989), pp. 7–24. Pierre Nora (ed.), *Les Lieux de mémoire*, 3 vols. (Paris: Gallimard, 1997).

[35] See Marc Jeck, 'D'Villa Louvigny', in Sonja Kmec, Benoît Majerus, Michel Margue & Pit Péporté (eds.), *Lieux de mémoire au Luxembourg. Erinnerungsorte in Luxemburg* (Luxembourg: Saint-Paul, 2007), pp. 209–214.

3 The Radio Sets

Radio sets constitute another key element of the radio *dispositif*, as they are objects allowing listeners to tune in to stations, making them the direct counterpart of the transmitters. As technical objects, their history had an important impact on radio experience overall, and the consumption of the medium throughout the twentieth century. Early radio sets were usually made by skilled amateurs, using crystal sets and headphones,[36] thanks to techniques spread by a transnational community of amateurs, from the 1910s onwards.[37] They became more common household items in the following decades, often occupying a spot in sitting rooms and kitchens, where they favoured a layout for family and group listening. The key post-war invention was the transistor set, and the first mobile pocket-sized radio set was developed in the United States in 1954[38]; however, this technology was only slowly adopted. According to Susan Douglas, the adoption of the transistor set developed youth culture in America, from the late 1950s, notably by the creation of new spaces of listening, due to the item's mobility: 'At work, in the car, on the beach, people—especially the young—brought radio with them and used it to stake out their social space by blanketing a particular area with their music, their sportscasts, their announcers'.[39] A similar phenomenon occurred in Europe, where sales of transistor sets increased throughout the 1960s.[40] This evolution of the radio receiver created new spaces of listening,[41] resulting in changes in the *dispositif*. Commercial radio stations seem to

[36] Chignell, *Key Concepts in Radio Studies*, pp. 97–99.

[37] Maria Rikitianskaia, "A transnational approach to radio amateurism in the 1910s", in G. Föllmer and A. Badenoch (eds.), *Transnationalizing Radio Research: New Approaches to an Old Medium* (Bielefeld: Transcript, 2018), pp. 133–140.

[38] S. L. Ellis, "Transistor Radios", in Sterling, *The Museum of Broadcast Communications*, pp. 1413–1414.

[39] Susan J. Douglas, *Listening In: Radio and the American Imagination* (Minneapolis: University of Minnesota Press, 2004, 1st edition 1999), p. 221.

[40] For more details about the history of the transistor set in France, see Elvina Fesneau, *Le Poste à transistors à la conquête de la France: La Radio nomade (1954–1970)* (Paris: INA Editions, 2011).

[41] These new spaces were mobile, as the transistor set could be easily carried by listeners. It also favoured both individual and group listening, giving the possibility for younger family members to listen to their own programmes on their own sets. It is important to note that these new spaces did not replace older ones, centred around family listening.

have welcomed these changes, as magazines such as *Fabulous* and *Salut les Copains* regularly published adverts promoting new models of transistor sets in their pages. Even if the magazines were consumer goods on their own, they also served as platforms for the advertisement of a multitude of other products (from beauty creams to clothes, by way of record players and transistors). These adverts were present throughout the magazines, sometimes even on full pages, rooting them into the materiality of the publications. As such, they were clear statements regarding the ways *Salut les Copains* and *Fabulous* were financed. Moreover, some adverts, in particular those selling transistor sets and other radio sets, played another role in the *dispositif*. By promoting specific items linked to radio listening and music consumption, they promoted goods which were playing a role in the material dimension of the *dispositif*.

The space of reception was not only shaped by the radio set, but also by the relationship between the item and listeners. Radio, as a medium, is partly defined by its potential 'secondariness', the idea that programmes can be listened to while conducting another activity, such as driving, cooking, and working. Although, this is in no way saying that radio ceased to be the centre of attention altogether,[42] simply that the medium allowed the possibility to do something else than give it undivided attention, making it, in the words of Michele Hilmes '[keep] us company during the rest of the day in our kitchens, bedroom, bathrooms, automobiles, offices, and workshops'.[43] While some broadcasters tried to fight this phenomenon at first,[44] it became gradually more accepted, also due to the idea that a less attentive audience would be more vulnerable to advertising.[45] This secondariness of radio echoes the difference between glance and gaze established by Ellis to differentiate between types of attention

[42] Chignell, *Key Concepts in Radio Studies*, pp. 99–102.

[43] Michele Hilmes and Jason Loviglio (eds.), *Radio Reader: Essays in the Cultural History of Radio* (New York: Routledge, 2002), p. 1.

[44] The BBC also tried to have an impact on the space of listening, although not on the physical layout, but the attention listeners put to the radio set. Indeed, according to Scannell and Cardiff, in the 1930s, the BBC encouraged listeners not to 'listen with half an ear' but to concentrate on the programmes, therefore trying to shape the experience of listening and the relation to the radio set. See Paddy Scannell & David Cardiff, *A Social History of British Broadcasting, Volume 1: 1922–1939, Serving the Nation* (Oxford: Blackwell, 1991), p. 370.

[45] David Hendy, *Radio in the Global Age* (London: Polity Press, 2000), p. 182.

given to cinema and television,[46] which also applies to other media, such as radio.[47] It is particularly important to remember that listeners could give more or less attention to radio, understood both as a medium and as object, as it impacts the relations between the constitutive elements of the space of reception.

It can be argued that the stations also attempted, at times, to have a more direct impact on the *dispositif* of reception, that is to say, on the spatial layout of the radio experience. One example can be found with the English service of Radio Luxembourg which is said to have created the 'Under the Bedclothes Club'. Few details can be found about this so-called club, however, the phrase 'under the bedclothes' (or, at times, 'under the bed sheets' or 'bed covers') appears to be strongly linked to the station; at least in the memory of its listeners.[48] Keith Richards, in his memoirs, also remembers listening to the station this way.[49] From this anecdote, it seems that listening to Radio Luxembourg[50] under the sheets and cover of one's bed was a rather common practice in Ireland and the United Kingdom, either at home or in the dorms of boarding schools. The point was to encourage young listeners—children and teenagers—to listen to evening programmes without disturbing other people in the

[46] John Ellis, *Visible Fictions. Cinema, Television, Video* (London: Routledge, 1982) p. 116.

[47] Jostein Gripsrud, *Understanding Media Culture* (London: Arnold, 2002), p. 213.

[48] For instance, the biography of Australian radio host Barry Alldis by his daughter is titled *Under the Bedclothes.* Interestingly, the phrase 'under the bedclothes' appears very often on social networks (Facebook groups, comments on YouTube videos, etc.) when former listeners remember listening to the stations in the 1960s and the early 1970s. One example of that is found on the personal website of Dick Orringa [https://offri056.home.xs4all.nl/radioluxembourg.htm] (last consulted 23 March 2023), who created, in the late 1990s, a page dedicated to the English service of Radio Luxembourg, which included a section for comments. This section ran from 1998 to 2015, gathering over 800 entries, almost entirely from former listeners of 208. Among those, 36 mention a memory of listening to the station under their bedsheets.

[49] 'Radio Luxembourg was notoriously difficult to keep on station. I had a little aerial and walked round the room, holding the radio up to my ear and twisting the aerial. Trying to keep it down because I'd wake Mum and Dad up. If I could get the signal right, I could take the radio under the blankets on the bed and keep the aerial outside and twist it there. I'm supposed to be asleep; I'm supposed to be going to school in the morning'. Keith Richards and James Fox, *Life* (London: Weidenfeld, 2010), pp. 52–53.

[50] And maybe to other stations, although it seems that the practice became more common at first with Luxembourg.

room and to avoid being caught by authority figures. This was only possible with a mobile radio set, which was listened to under the covers, thus preventing listeners from being heard and seen. This likely created a sense of rebelliousness and trespassing in the young listeners, while also creating a peculiar social and cultural space of listening,[51] supported and encouraged by the commercial radio station. A second example of the stations shaping—or, at least, suggesting—a space of reception is found in the common show broadcast by RTL and Europe n°1 in May 1968.[52] In this programme, analysed in Chapter 7, two radio hosts explain to listeners how they should position two receivers, one for each station, in a V-shape layout, in order to fully experience the stereophonic joint programme. This example shows how stations could have an impact on the listening space, by suggesting an ideal layout, due to an exceptional change in the production space: the temporary collaboration of RTL and Europe n°1.

While the radio receiver as an object plays a central role in the *dispositif* of commercial radio stations, it is noteworthy to mention that the concept extends to smaller material elements, such as radio dials on receivers. The symbolic importance of the dial was already considered previously, where its role in the appropriation of an imagined European broadcast landscape was discussed. It can be argued that within the radio set, it is the dial, a small technical element of the receiver, which carries the symbolic dimension by its ability to evoke a transnational broadcasting space. Moreover, communication technologies have allowed listeners to 'travel' and have the feeling of belonging to national and transnational communities,[53] inferring a symbolic connection between the sets and their users. The crucial role of the dial in the *dispositif* of commercial radio stations is even more striking when, in 1971, manufacturers decided to add buttons for Radio Luxembourg on their radio sets destined for

[51] This new space of listening echoes the role of radio on the reconfiguration on the private and intimate space overall: "radio played an important role in reconfiguring the nature of the private, of intimate space—it was being integrated into individual lives, into individual private fantasies". Taylor, Katz and Grajeda, *Music, Sound, and Technology in America*, p. 247.

[52] Common radio programme, 8th May 1968, RTL & Europe n°1, 02:31:58, personal archives of Roland Biesen.

[53] Shaun Moores, *Interpreting Audiences: The Ethnography of Media Consumption* (London: Sage, 1993), p. 22.

Germany. This occurred after the improvement of the short-wave transmitter of the station and the subsequent impact on the listening quality of the German programmes.[54] As mentioned previously, the board of directors of Radio Luxembourg rejoiced at the news, which was seen as decisive to increase listenership and subsequent advertisement revenue, echoing the commercial nature of the broadcasters. Therefore, a close look at a single element—the radio dial on a set—reveals the extent of the *dispositif* even in small details, as well as the connections to commercial and symbolic dimensions.

This brief study of the radio set reveals the importance of this element within the *dispositif* of commercial radio stations, as it bridges listeners to the content. Therefore, it became the central element in shaping the space of reception, which the stations tried, at times, to control and influence. The set, especially through the dial, also carried a strong significance, bridging material, commercial, and symbolic dimensions of the *dispositif*.

4 Magazines & Other Commodities

While this chapter has so far focused on the broadcasting side, it is crucial not to neglect the commodities produced and sold as part of the commercial venture of the *dispositif*. Among them, the *Fabulous* and *Salut les Copains* magazines represent the most striking example. These publications, printed and distributed *en masse*,[55] also constituted a crucial aspect of the material dimension, notably due to their accessibility and mobility. Moreover, as the analysis of the pen pal sections in Chapter 8 revealed, the readership of a magazine like *Salut les Copains* was wide and international,[56] showing a potential reach beyond what the Europe n°1's transmitter could offer. As such, magazines constitute important

[54] Minutes of the 'Assemblée générale annuelle', 18th May 1971, RTL Group Archives, Luxembourg.

[55] *Salut les Copains* sold particularly well, with slightly over 180,000 copies with the first issue, and around a million in 1963, while *Fabulous* sold around 300,000 per issue. See Christophe Quillien, *Nos Années "Salut les Copains": 1959–1976* (Paris: Flammarion, 2009), p. 66, & Chris Tinker, "Shaping 1960s youth in Britain and France: Fabulous and *Salut les Copains*", *International journal of Cultural Studies* (14:6, 2011), pp. 641–657.

[56] For instance, *Salut les Copains* published letters from readers living, for instance, in Spain and in England, while the pen pal section revealed the presence of readers in Vietnam and Canada for instance.

elements of the *dispositif*, as they incarnate intermedial bridges, essential components of the 'plurimedia offer' described by Claire Blandin.[57]

Among the vast network of intertextual and intermedial references operating between the radio stations and the magazines, one kind of reference is of particular interest here. The *Fabulous* weekly publication of the English service of Radio Luxembourg's programme listings represents a systemic structural component which supports the promotion of the station's programmes in the pages of the magazine. The idea of radio shaping the temporal rhythm of daily lives,[58] a phenomenon reinforced by the magazine's publication of radio listings, was already discussed. As the programmes can also be understood as commercial commodities, this points to the coordination in production between *Fabulous* and Radio Luxembourg as part of a wider system in which both media support one another.[59] The specificity of magazines can be found in the importance and attention given to images and photos, in addition to its textual elements, such as articles, editorials, or letters. This allowed a magazine such as *Salut les Copains* to 'show people what they heard'[60] on the eponymous radio programme, although it could be argued that a similar logic was operating with *Fabulous 208*. At times, this can be understood in an almost literal way, as magazines published pictures taken in the studios for instance. More generally, magazines operated as a platform to support a variety of norms,[61] discourses and especially images and values also present in the radio shows. These images and values, called 'features' in a previous chapter, constitute another example of the symbolic dimension through which the magazines furthered shared characteristics of the stations. More precisely, the intermedial features were identified as (1)

[57] Claire Blandin, "Radio et magazine: une offre plurimédia pour les jeunes des Sixties", *Le Temps des Médias* (2–21, 2013), pp. 134–142.

[58] Indeed, as explained by Gripsrud: 'broadcast media form a 'normal' time scheme for 'normal' lives, they structure each day and give each day of the week a special character'. Gripsrud, *Understanding Media Culture*, p. 24.

[59] This is also a way to detach the analysis from an essentialist reading which would limit Europe n°1 and Radio Luxembourg as 'only' radio stations and their programmes.

[60] Franck Ferrand, "Au coeur de l'émission Salut les Copains!", *Au Coeur de l'Histoire*, Europe 1, 14th September 2012, with Daniel Filipacchi, Hubert Wayaffe and Eddy Mitchell.

[61] Such as the name of *copains* used to define listeners, readers, and celebrities linked to *Salut les Copains*.

irreverence and playfulness, (2) commercialism, (3) localism and transnationalism, and (4) interactivity. This echoes the aforementioned definition of the *dispositif* made by Giorgio Agamben, for whom the concept has 'the capacity to capture, guide, determine, intercept, shape, control, and ensure the gestures, behaviors, opinions and discourses of living beings',[62] something such shared features in both media could do. The question of commercialism in both media is slightly different, due to the presence of an external group of actors: advertisers. To use Michael Stamm's phrasing, they worked with 'space in print and time in sound',[63] suggesting their versatility, and presence in both media.

The mention of advertisers echoes the point made previously that magazines are part of a wider group of commercial commodities within the *dispositif*. Some of these commodities are directly produced by commercial radio stations, such as magazines and radio programmes. It could be argued that other commodities are promoted through the *dispositif*. Among them, products (*e.g.* record players, cigarettes, shaving creams) advertised on air and in the pages of *Fabulous* and *Salut les Copains* were pushed forwards by said advertisers and resulted in the main revenue stream of the *dispositif*. Commodities also included various celebrities. For instance, the *Photo du siècle* portrays several artists linked to *Salut les Copains* (the magazine and the radio show), while the *A-Z of Fab Fancies* draws a list of 400 names of most-known celebrities; according to *Fabulous 208* and including some of the disc-jockeys of Radio Luxembourg. Therefore, it appears that the *dispositif* produced and supported numerous commodities, whether directly or not, which all belonged to some extent to the overall commercial purpose. The analysis conducted in chapters 8 and 9 revealed the strength of the intermedial references in both radio and magazines, and how they were echoing one another. Nevertheless, it does not mean that they were designed to be commodities consumed—or, rather, listened to and read—together, notably because they operated on different temporalities. One example, however, found in a recording of *Salut les Copains*, displayed such a joint consumption, and even encouraged it for listeners. In this source from

[62] Agamben, *Qu'est-ce qu'un dispositif* ?, p. 30.

[63] Michael Stamm, "The Sound of Print: Newspapers and the Public Promotion of Early Radio Broadcasting in the United States", in D. Suisman & S. Strasser (eds.), *Sound in the Age of Mechanical Reproduction* (Philadelphia: University of Pennsylvania Press, 2010), pp. 221–241.

April 1965, French singer Claude François acts as the host of the Europe n°1's programme, although he clearly appears out of his comfort zone.[64] During the programme, he talks about his activities, and details his latest trip to England and to the United States. What is particularly interesting, however, is the fact that during the American part of his journey, Claude François was accompanied by Jean-Marie Périer, the lead photographer of the *Salut les Copains* magazine. Périer took many shots of the singer, which were published in the latest issue of the magazine, an issue that Claude François had in front of him while presenting the radio show, and could be heard flipping through on air. The singer shared many anecdotes and comments, all linked to the photos published, leading him going back and forth through the magazine to find the photo he is referring to, sometimes even precising the page.[65] This recording displays how a rather unique example of the two media being strongly interlocked, as the magazine is read on air,[66] during the show. Moreover, the singer's comments, especially when he points at a specific page, can be read as a discourse suggesting listeners to be readers at the same time. By doing so, he was encouraging them to follow his exploration/consultation of the latest issue of *Salut les Copains*, hinting at a space of reception using both a radio set and a magazine.

One example developed by Barthes in his work on myths echoes the research undertaken here. On regular occasions in his demonstration, Barthes mentions the cover of an issue of a French magazine. On it, a young black soldier in French uniform saluting; an image used by Barthes to showcase his methodology and deconstruct the image's symbolism and meaning. What is of interest, however, is that the magazine in question was *Paris Match*, a major news magazine in France founded in the late 1930s by Jean Prouvost. The same Jean Prouvost who took over the French service of Radio Luxembourg in 1966, and launched a series of changes, including the rebranding of the station to RTL. *Paris Match* is also the magazine which was bought in 1976 by Daniel Filipacchi,

[64] The singer indeed shows little enthusiasm, hesitates, pauses, repeats himself, etc. on numerous occasions. Salut les Copains, présenté par Claude François, April 1965, 00:39:21, Europe 1 archives, Paris.

[65] For instance, Claude François explains that, on page 103, he can be seen posing next to a 'big red car'.

[66] As indicated by both Claude François' comments and the clear sounds of pages being turned.

host and editor of, respectively, the radio show and the magazine *Salut les Copains*.[67] These elements are more than mere coincidences, as both Prouvost and Filipacchi had a large impact on their respective media and their editorial content due to their positions. This shows that—in the French context—the ties between magazines and commercial radio were also the result of a handful of men such as Prouvost and Filipacchi, who had interest and influence in both media. As a result, they can also be considered as influential elements of the *dispositif*, who mostly had an impact on the dimensions of commercialism and content.

Magazines, understood as elements of the *dispositif* of commercial radio stations, differ from the others for their non-essential nature. Indeed, magazines are not needed for radio broadcasting, in the way transmitters and receivers are; however, they are still defining elements. They have a particularly strong impact in shaping the stations' and the programmes' images, notably through intermedial shared features. Interestingly, their non-essential nature was echoed in their temporary participation to the *dispositif*. *Salut les Copains* was first published in 1962, and its relation to Europe n°1 weakened after the end of the eponymous radio show in 1969, before the relationship was severed in 1974 when the magazine was renamed *Salut*, a clear step to distance itself from radio. Similarly, *Fabulous* was first published in 1964, but only partnered with Radio Luxembourg in 1966, and was cancelled in 1980. This example clearly shows a beginning and an end for this element of *dispositif*, one which could be added to and removed from the wider infrastructure,[68] and subject to changes over time. It is not possible to list all the commodities produced and sold by the *dispositif*, however, by focusing on magazines, this section sheds light on a crucial example, which reveals a vast range of intermedial interrelations. Moreover, magazines themselves acted as vessels to promote other commodities, reinforcing their relevance here.

[67] And of *Mademoiselle Âge Tendre* and other magazines, making him a media tycoon.

[68] A similar dynamic can be found with the short-lived *Feu de Camp* programme, which produced a new transmedial element, quickly cancelled after it did not produce the expected results.

5 The Sound

The fifth constitutive element of the *dispositif* is the sound of commercial radio stations. The research conducted for this book was based on roughly 80 hours of recordings of commercial radio, and it is this ensemble which is considered here as an element of the *dispositif*. As such, the sound is more than a succession of independent elements, and radio programmes are understood as part of a flow. Moreover, the study of a large corpus of audio sources reveals a series of key features, which define the sound of commercial radio stations. An approach which would favour a more fragmented vision of the programmes could miss some of the power dynamics at stake in the *dispositif*.[69] The sound represents a key facet of the wider apparatus of the stations, as it connects the radio stations and their programmes to listeners, who have an active role in it. And part of this active role is linked to the 'the pre-expressive ear of the listener',[70] hinting at the necessary knowledge and experience listeners had, in order to interpret—or decode[71]—radio texts and sonic elements present in the sound. Andrew Crisell, in *Understanding Radio*, borrows from semiotics and suggests three categories of auditory signs to be analysed in radio: words,[72] sounds, and music.[73] The three categories are of interest here. They are produced by broadcasters and radio hosts, using communication technologies and radio norms, have an impact on stations' images, and they are meant to be decoded by listeners. This logically brings the sound to the centre of various dimensions (*e.g.* material, imagined) of the *dispositif*, which echoes the idea that 'production, representation

[69] Moreover, as the amateur recordings studied in Chapter 6 seem to indicate, listeners did not tune in and out at the beginning and end of a programme, reinforcing the necessity to study them as a whole.

[70] The active role of the listener in a radio soundscape was mentioned, for instance, by Frances Gray, who carried research on radio plays. Frances Gray, "Fireside Issues: Audience, Listener, Soundscape", in Andrew Crisell (ed.), *More than a Music Box: Radio Cultures and Communities in a Multi-Media World*, 2nd ed. (New York: Berghahn Books, 2006), pp. 247–262.

[71] The idea that popular culture can be read as a socially constructed system of codes and signifiers was central in the work of the Centre for Contemporary Cultural Studies at the University of Birmingham. See Stuart Hall, 'Encoding/decoding', in S. Hall, D. Hobson, A. Lowe and P. Willis (eds.), *Culture, Media, Language: Working Papers in Cultural Studies 1972–1979* (London: Taylor & Francis e-Library, 2005), pp. 117–127.

[72] And, by extension, the various voices heard on radio speaking them.

[73] Andrew Crisell, *Understanding Radio*, 2nd ed. (London: Routledge, 1994), p. 42.

and consumption do not exist as autonomous spheres',[74] an underlying characteristic throughout this chapter.

The human voice is undeniably a crucial element of the radio sound and, according to Smati and Fiévet, even incarnates the *dispositif radiophonique*.[75] The voice serves multiple purposes in radio, and 'words on the radio are always and unavoidably *spoken*',[76] meaning that they are a binary code which has to be read alongside the voice in which they are heard. And this voice carries meaning depending on many factors, such as the person or character who speaks, accents, and intonations,[77] something well known since the early days of radio.[78] Moreover, voices on radio can act as unifiers of programmes, bringing together the various sections through one host or disc-jockey, while voices of announcers often identify the whole station.[79] Whether it was on Europe n°1, or on the English or French service of Radio Luxembourg, female voices were seldom heard on commercial radio. Women were often limited to being announcers and to reading adverts, as Françoise Durgis in *Feu de Camp* did on Europe n°1 for instance. These roles are important, especially with commercial broadcasters, but as on-air personalities, their voices are in the background. There is, however, one exception. Menie Grégoire[80] was a particularly well-known voice of RTL, where she worked over a decade, hosting famous programmes such as *Allô, Menie?* and *Responsabilité sexuelle*. In these programmes, Grégoire discussed questions of intimacy, sexuality, and marital life, often sparking controversy.[81] In addition to the critics about the use of psychology and psychoanalysis in her programme, she

[74] Bill Osgerby, *Youth Media* (London: Routledge, 2004), p. 12.

[75] Nozha Smati and Anne-Caroline Fiévet, "A la radio, la voix donne à écouter et à voir", 4th December 2017, INAglobal.fr, 5 pages.

[76] Crisell, *Understanding Radio*, p. 43.

[77] William O'Donnell and Loreto Todd, *Variety in Contemporary English* (London: George Allen & Unwin, 1980), p. 95.

[78] See Tom H. Pear, *Voice and personality* (London: Chapman and Hall, 1931).

[79] Crisell, *Understanding Radio*, p. 43.

[80] Marie 'Menie' Grégoire (1919–2014), was a French writer and journalist, mostly known for her career on RTL, where she worked from 1967 to 1981.

[81] Denis Maréchal, *RTL. Histoire d'une radio populaire* (Paris: Nouveau Monde, 2010), pp. 339–362.

was criticised for supporting a form of voyeurism and sensationalism.[82] Despite critics and controversies, her programmes had considerable listenership.[83] Not only was she a long-lasting female voice on a prominent commercial station, she gave a voice to many women, either directly via phone-ins or, most of the time, indirectly, by reading and answering on-air letters that she received.[84] Nevertheless, Menie Grégoire represents more of an exception for female voices within a field dominated by male personalities.

Voices and spoken words give a 'flavour' to the sound of commercial radio. For instance, in *Pour ceux qui aiment le jazz*, it is through the way they address each other—the tone of the voices and word choices—that the hosts Ténot and Filipacchi create a sense of intimacy, which listeners are invited to join.[85] Voice is also the main tool used by radio hosts and disc-jockeys to shape their image and on-air personalities. In this regard, the detailed description made of one of Kenny Everett's shows on Radio Luxembourg[86] highlights the importance of its voice to create a form of irreverence. The British disc-jockey indeed uses a wide range of vocal techniques, such as juggling between high and low pitches, making sexual sounds, and imitating a Scottish accent—to name a few examples—to develop this irreverent image. Such behaviours have an impact on the station's image itself, and not only on specific personalities. Keeping in mind the context of the *dispositif*, it is important not to forget that the interdependencies between voices and radio stations also occurred from a 'top-down' perspective. Indeed, a station like RTL gave guidelines to shape this image. For instance, through a memo reminding journalists that the station wished to display a European image, and that they had

[82] Anna Jehle, *Welle der Konsumgesellschaft: Radio Luxembourg in Frankreich 1045–1975* (Göttingen: Wallstein, 2017), pp. 146–155.

[83] Denis Maréchal quotes an average of 1,3 million daily listeners, and 40% of total listenership in France. Broadcast in the early afternoon, the programme targeted and was mostly listened to by women. Maréchal, *RTL. Histoire d'une radio populaire*, p. 343.

[84] Menie Grégoire shares, in her memoirs, several extracts of the letters she received during her radio career (over 100,000 letters in roughly fifteen years). See Menie Grégoire, *Comme une lame de fond* (Paris: Calmann-Lévy, 2007).

[85] Pour ceux qui aiment le jazz, Europe n°1, 1955, 00:20:30, Archives Europe n°1, Paris.

[86] Esso Show, presented by Kenny Everett, Radio Luxembourg 208, 13th November 1970, 00:29:07, personal archives of Mick Capewell.

to keep this in mind when on air.[87] This shows how the voice, a central element of the sound, could shape a station's image, as well as how a station had a hand in the direction taken by such an image.

Irreverence and playfulness were some of the key features defined after the lengthy study of the sound of commercial radio stations. These characteristics of the sound, which is studied here as an element of the *dispositif* of commercial radio stations, highlight how they stood out from public broadcasters. Moreover, the focus on light entertainment for instance, as well as the presence of advertisements, played in this distinction. In this context, the commercial nature of the stations was melted into their sound, which can 'manipulate mood, especially as it relates to the purchasing of consumer goods that one might think are unrelated to sound'.[88] This reveals the recurring presence of commercialism and advertising in many facets of the studied *dispositif*, for it is its strategic function. Phone-ins are quite specific to the medium of radio, and commercial radio also made use of the technique, which 'enables radio broadcasters to create the illusion of a two-way medium and to verify both that they have an audience and the audience is capable of responding to the codes they transmit'.[89] This example points to the use of technology, incorporated into the sound that had a strong impact on listeners.

While the voice holds a key place in radio, it coexists alongside various sounds. They constitute central elements of the sound of commercial radio stations, as shown previously in this book. Moreover, sounds and noises do not exist in radio as they do in everyday lives. On air, sounds are carefully chosen and created for listeners: 'radio does not seek to reproduce the chaotic, complex, and continuous sounds of actual life [...] but seeks to convey only those sounds which are relevant to its messages and to arrange them in a hierarchy of relevance. [...] Broadcasters must prioritize sounds for [listeners], foregrounding the most

[87] Memorandum from Michel Moine to Jean Le Duc, 10 December 1964, RTL Group Archives, Luxembourg.

[88] As mentioned by Goodale, himself basing his comments on the research undertaken by Sterne. See Greg Goodale, *Sonic Persuasion: Reading Sound in the Recorded Age* (Urbana: University of Illinois Press, 2011), p. 144, and Jonathan Sterne, 'Sounds like the Mall of America: Programmed Music and the Architectonics of Commercial Space', *Ethnomusicology* (41:1, 1997), pp. 22–50.

[89] Crisell, *Understanding Radio*, p. 61.

important ones and eliminating the irrelevant ones'.[90] The hierarchy of sounds is for instance displayed by the example of an interview, in which the recorder is set in a way to put voices in the foreground.[91] Similarly, studios, with their specific acoustics, shaped the hierarchy of sounds, and, as mentioned previously, reinforced the intimacy with listeners. In both cases, the space of production and technologies are used to produce and hierarchise sounds, inferring the interdependencies between the sound and the material dimension.

Among the various sounds of commercial radio stations, some held a particularly strong significance.[92] Sonic icons, such as jingles, were fundamental. Examples developed previously, such as the jingle composed by Michel Legrand for Radio Luxembourg, and those made for *Salut les Copains* with famous singers of the period, show how much effort the stations put in such sounds. Sonic icons had an important role, as they helped identify stations and pointed to key moments within a programme. Moreover, the use of celebrities can be seen as a display of strength, and helped shape specific elements, such as the association between stations and addresses, developed previously. In every case, sonic icons are the result of careful and attentive production, had an impact on various

[90] Crisell, *Understanding Radio*, pp. 44–45.

[91] For more details about this example, see Crisell, *Understanding Radio*, p. 45. In other settings, technology might capture sounds which accidentally reinforce the message made by a reporter. For instance, this was the case during the protests of May 1968 in France, when a reporter got caught in the middle of an altercation between protests and police forces. On the recording, his microphone caught a series of sounds (a stone hitting him, the sounds of boots running on the pavement, etc.) which created an 'effect of reality' on listeners. See Richard Legay, "RTL & Europe n°1 as central actors. The Importance of Mobility for Commercial Radio Stations during the Parisian Events of Mai 68", *Rundfunk und Geschichte* (3:4: 2018), pp. 41–51.

[92] Music was highly present in the 'sound' of commercial radio stations in the 1960s, however, its study in the context of this book presents challenges, despite the interesting connections between the stations and the record industry. The use of music as sonic illustrations in a radio play for instance is fairly easy to interpret, however, in the case of Europe n°1 and Radio Luxembourg, music is often played on its own, as 'an object of aesthetic pleasure in its own right', which is not modified by radio, and therefore does not illuminate it. See Crisell, *Understanding Radio*, p. 49. Furthermore, scholars have mentioned that music now is made to be heard rather than listened to, due to the 'machinic precision that turns music into the sound of mass production and consumption' Jacques Attali, *Noise: The Political Economy of Music*, 12th ed. (Minneapolis: University of Minneapolis Press, 1992), p. 4.

dimensions of the *dispositif*, and were made with listeners in mind.[93] As an element of the *dispositif*, the sound is important as it is in direct contact with listeners and it is particularly rich. Born out of the material dimension, it is more than just content, as it carries meaning through voices and sounds. Moreover, the study of the sound shows how listeners held a relatively active role in the entire *dispositif*, for they had the tools and knowledge to interpret and decode the various auditory elements.

6 The Myth of Commercial Radio

Through their radio shows, especially of light entertainment, and their partner magazines, commercial radio stations shared more than simply entertaining content, they shared a vision of popular culture, an ensemble of commodities, a celebrity culture, and, last but not least, a set of values. In this regard, the works of Tinker and Pires are particularly insightful to understand the construction of masculinities,[94] notions of youth,[95] and representations of Black artists.[96] This ensemble of representations and values evolved, to some extent, throughout the period, although it retained homogeneous characteristics. Such an ensemble echoes directly what Roland Barthes coined through his concept of myth,[97] which is

[93] A few media scholars have mentioned the importance of redundancy in radio. Due to its ephemeral nature, many signs are repeated and made of predictable and conventional material. Crisell, *Understanding Radio*, p. 59; Walter Ong & John Hartley, *Orality and Literacy*, 3rd ed. (London: Routledge, 2013), pp. 39–40; John Fiske & Henry Jenkins, *An Introduction to Communication Studies*, 3rd ed. (London: Routledge, 2011), pp. 11–13.

[94] Chris Tinker, "Mixed Masculinities in 1960s British and French Youth Magazines", *The Journal of Popular Culture* (7:1, 2014), pp. 84–108.

[95] Tinker, "Shaping 1960s youth in Britain and France", pp. 641–657.

[96] Mat Pires, "Les stars noires et *Salut les Copains*, 1962–1968", *Communication et langages* (n°111, 1997), pp. 59–71.

[97] French literary theorist and philosopher Roland Barthes published his renowned book *Mythologies* in 1957. In it, he re-published, in the first part of the monograph, a series of essays looking into what he calls modern myths, which are created by contemporary society. In the second part, entitled 'Myth Today', Barthes deconstructs his vision of the concept of myth, which highlights the connections between speech and power, operating as a way to naturalise particular worldviews, which applies well to the set of values described above. Roland Barthes, *Mythologies*, revised ed. (London: Vintage, 2009, translation by Jonathan Cape). In 2007, a book, containing 'new myths' was published to commemorate the 50th anniversary of Barthes' *Mythologies*. See Jérôme Garcin (ed.), *Nouvelles mythologies* (Paris: Seuil, 2007).

another element of the *dispositif*, one spread on both dimensions of content and significance.

A myth is 'a type of speech, [...] a system of communication [and] it is a mode of signification', and, therefore, as a type of speech: 'everything can be a myth provided it is conveyed by a discourse'. Furthermore, this means that a 'myth is not defined by the object of its message, but by the way in which it utters this message',[98] making it a particularly useful tool and vessel to encompass the 'message' spread by commercial radio stations. Indeed, Barthes' definition of the myth makes it a general reading of the world, not simply linguistics matters.[99] Moreover, 'speech' should not be understood simply as oral discourse, for it covers 'modes of writing or of representations; not only written discourse, but also photography, cinema, reporting, sport, shows, publicity, all these can serve as a support to mythical speech'.[100] Such a vast range of support directly echoes the questions of intermediality and media entanglements raised previously in this book. The concept of myth, in Barthes' vision, works in a way as an extension of semiology, notably through the works of Saussure.[101] The mythological system builds on this relationship to add another sign, which is elevated to a myth. Moreover, 'the materials of mythical speech (*e.g.* the language itself, photography, painting, posters, rituals, objects), however different at the start, are reduced to a pure signifying function as soon as they are caught by myth'.[102] The use of the concept of myth here is the ideal way to show how the content produced by the *dispositif* of commercial radio stations also carried many supplemental meanings.

The power of the myth is to offer a representation of the world which incorporates an ideology, a specific vision of reality, through a form of

98 Barthes, *Mythologies*, p. 131.

99 Umberto Eco and Isabella Pezzini, "La Sémiologie des *Mythologies*", *Communications* (36, 1982), pp. 19–42.

100 Barthes, *Mythologies*, p. 132.

101 In a few words, this discipline analyses the connections between an object—the signified—and its linguistic representations—the signifier—and how to create the third element: a sign. A clear example of that can be found in the rose. Described in *Mythologies*, it goes as follows: 'It is as true to say that one the plane of experience I cannot dissociate the roses from the message they carry, as to say that on the plane of analysis I cannot confuse the roses as signifier and the roses as sign: the signifier is empty, the sign is full, it is a meaning'. Barthes, *Mythologies*, p. 136.

102 Ibid., p. 137.

repurposing. In the topic at hand, the aforementioned ideology is directly inspired by capitalism and consumerism. An example of repurposing is found in songs regularly played on Europe n°1 and Radio Luxembourg, and often advertised and spoken about in the magazines, see their original artistic value, the meaning of the lyrics vanish, to be turned into commercial commodities, and 'things to know' (such as in the *A-Z* analysed previously) as part of a wider popular culture. And that is where the power of the myth lies, as '*myth hides nothing*: its function is to distort, not to make disappear'.[103] This distortion is one of the most fascinating aspects of the concept of the myth, especially because a myth is consumed innocently. Indeed, a myth presents itself as a system of facts, and is consumed as such, whereas it is actually a semiological one. Therefore, it can be deconstructed in order to be understood, which this section attempts to do in the following paragraphs. In addition to the links between intermediality and myths, intertextuality plays a role as well. Indeed, 'mythical speech is made of a material which has *already* been worked on so as to make it suitable for communication: it is because all the materials of myth (whether pictorial or written) presuppose a signifying consciousness'.[104] In other words, every radio show, every jingle and sonic icon, every magazine, and photograph are all connected through mythical speech. And this is part of the unusual nature of the concept, whose 'charm' can be described as a *pêle-mêle* (a jumble), which mixes American-Swedish actress Greta Garbo and chips; Citroën's famous 'D.S.' car and well-known French priest *Abbé Pierre*.[105] This echoes the empirical analysis done in this book so far, in which a jingle rubbed shoulders with a map, minutes of a meeting with a letter published in a magazine, and so forth. This points to a crucial limit to the concept of myth, as understood by Barthes. Indeed, in his work, myths appear as fragmented and discontinuous, and do not offer a clear narrative, meaning they cannot be told, they can only be analysed.[106] Therefore, this means that the study of the contemporary myth of commercial radio stations can only act as a step

[103] Barthes, *Mythologies*, p. 145.

[104] Ibid., p. 133.

[105] Philippe Roger, *Roland Barthes, roman*, 2nd ed. (Paris: Grasset, 1990), p. 94.

[106] Which also shows how the scholarly tradition on myths was discarded by Barthes. See Serge Zenkine, "L'Esthétique du mythe et la dialectique du signe chez Roland Barthes", *Littérature* (108, 1997), pp. 102–124.

towards a more final understanding of the topic at hand,[107] namely, the production apparatus behind the myth; the *dispositif* of commercial radio.

The first question to be asked regarding the 'demystification' of commercial radio stations is to identify where the myths can be found. As mentioned previously, any type of discourse can support mythical speech, therefore, it makes sense to look for it in the large variety of commodities produced by commercial radio stations. In the case of radio programmes, previous analysis showed that commercialism was a key feature of the sound of Radio Luxembourg and Europe n°1. Moreover, adverts were not separated from the rest of the programmes, they were blended into the flow of radio, reinforcing the transformation of 'one culture's values [...] into a universal and natural value'[108] operated by mythology. Indeed, through the careful integration of adverts into their sound, radio stations like Radio Luxembourg and Europe n°1 did not hide their commercial nature, but rather turned it into something normal and natural.

The importance of photography for magazines, discussed previously, is also relevant here. As mentioned by Barthes, 'pictures become a kind of writing as soon as they are meaningful',[109] therefore, magazines such as *Salut les Copains* and *Fabulous 208*, which put a great deal of effort in their photos, become powerful vessels for the mythical speech developed by commercial radio stations. Photos picturing disc-jockeys from the English service of Radio Luxembourg in silly and absurd positions after horse-riding lessons can be read as a promotion of their sense of irreverence. Interestingly, it is necessary for the 'myth-reader' of these pictures to be aware of this type of photographic representation to understand its parodic nature in the magazine, as 'mythical speech is made of a material which has *already* been worked on',[110] indicating, once again, the

[107] This is also due to the fact that Barthes never really explains how to 'hunt' for myths, and only seems to be doing so by instinct (See Serge Zenkine, "Les Indices du mythe", *Recherches & Travaux* (77, 2010), pp. 21–32). Therefore, a systematic analysis of the myths produced by commercial radio stations would not be a doable enterprise, and is limited to a handful of examples; following the example set by Barthes himself in the first part of *Mythologies*.

[108] Graham Allen, *Roland Barthes*, 2nd ed. (London: Routledge, 2004), p. 37.

[109] Barthes, *Mythologies*, p. 133.

[110] Barthes, *Mythologies*, p. 133. This also echoes the aforementioned work of the Birmingham school and scholars such as Stuart Hall.

importance of intertextuality. Photos played a key role in the naturalisation process of stardom and celebrity culture in this period. Furthermore, as was shown previously, they furthered the inclusion of radio personalities into a celebrity culture, a process which started a few decades before.[111] By doing so, disc-jockeys and radio hosts appeared as naturally belonging to this circle of happy few, itself a social construct. This points to the importance of remembering that mythical speech is not limited to the context of commercial radio stations and their magazines. Indeed, the myth of stardom already pointed at a larger phenomenon, while the representations of black celebrities in *Salut les Copains*, studied by Mat Pires,[112] showed how photos and photo-reportages furthered colonial representations in popular culture. Both these examples indeed echo the wider context of popular culture in Western Europe in the Long Sixties. These links between mythical speech and the wider popular culture and contemporary values make it stand out as an element of the *dispositif*, as evolutions in popular culture impacted the myth, which in return impacted popular culture.[113]

7 The Receiving End

The seventh and last element of the *dispositif* analysed here is the audience, the people listening to the stations and reading the magazines, those destined to receive the product of the entire apparatus. They are not understood in this section so much as actual people, but rather as their interpretation and construction by the stations.[114] For instance, their representation on air and in letters published in *Salut les Copains*, or as numbers in listernships surveys. Their definition therefore presents some challenges as they are all at once listeners, readers, consumers, and users,

[111] See, for instance, *Das Buch der Ansager: Die Ständigen Rundfunkansager der Europäischen und der Grossen Amerikanischen Rundfunkgesellschaften in Wort und Bild* (Berlin: Rothgiesser & Diesing A.G., 1932), which portrays many radio hosts of the time.

[112] Pires, "Les stars noires et *Salut les Copains*, 1962–1968", pp. 59–71.

[113] For instance, new popular records would be played by the station, while the station's promotion of a new record would make it more popular.

[114] Although, it is important that in many cases, radio presenters address, on air, listeners as an individual, which creates tensions with the perception of the audience as a consuming mass. Chignell, *Key Concepts in Radio Studies*, pp. 64–65.

as well as members of a wider community, depending on the perspective chosen. While the *dispositif* of commercial radio stations is wired in a way to support their business-oriented nature, it is also oriented towards the audience.[115] Furthermore, everyone involved in the process of making radio reflects on who listens,[116] and this subsequently impacts the content, something especially salient for commercial radio stations. For them, the larger the listenership the higher the income, thus creating a need to cater to listeners' interests and wishes. It is, however, not easy to clearly capture the number and the nature of listeners, for some might record radio programmes to listen to them later. Some might even listen to recordings multiple times, particularly when they record radio for the songs played.[117] This first point shows the place of listeners in the *dispositif*, at the receiving end, and how, because of this position, they have an impact on multiple elements, especially the content, inferring their active role in the ensemble.

Due to the strong intermedial connections, notably with magazines, people at the receiving end cannot be solely defined as listeners, but also as readers. These two identities are not, however, irreconcilable as many people were both. In a similar way, listeners of Europe n°1 could also very well be listeners of Radio Luxembourg. Keeping in mind the commercial nature of the subjects of this book, the listeners-readers were also seen as consumers, and, through that lens, turned into something measurable. This phenomenon is not limited to radio and the press, but the product of 'the distinct but overlapping developments of the recording industry, the movie industry, and broadcasting [which] all fueled the modern notion of mass audiences as great multitudes of customers'.[118] The analysis conducted previously on surveys illustrates rather clearly how listeners could be turned into numbers, and how crucial these numbers were for both Europe n°1 and Radio Luxembourg. Where maps become a visual representation of the technical dimension, listenership surveys

[115] Indeed, as mentioned by Chignell: 'all radio is made for an audience, for the mass of listeners'. Chignell, *Key Concepts in Radio Studies*, p. 63.

[116] Hendy, *Radio in the Global Age*, p. 115.

[117] 'The many who illegally record pop music from the radio in order to listen repeatedly to the hits they would otherwise have to buy', which might explain some of the origins of some of the sources mentioned in previous chapters. Crisell, *Understanding Radio*, p. 202.

[118] Taylor, Katz and Grajeda, *Music, Sound, and Technology in America*, p. 4.

are a textual reading of the commercial dimension through audiences. In regards to the *dispositif*, these surveys (about listeners in France, the stations' biggest market) first indicate evolutions and changes throughout the Long Sixties as numbers evolved. Second, surveys were conducted by external actors, not by the stations themselves,[119] which gave them a more neutral and objective nature. As such, these surveys became metrics for the stations, and the data they provided was used in the competition between commercial broadcasters. In a similar way, readers of magazines such as *Salut les Copains* and *Fabulous* also represented key interests, and external actors were hired to conduct surveys to measure readerships. In France, the C.E.S.P., the *Centre d'Etude des Supports de Publicité*, was one of the institutions in charge of conducting such surveys, destined to magazines and advertisers.[120] This was the same institution conducting surveys on radio listenership from 1968 onwards, reinforcing the closeness in nature of readers and listeners, at least in the minds of commercial stations and advertisers, for whom their role as consumers was fundamental.

One last point should be raised in regards to the status given to those 'at the receiving end' of the *dispositif* and how they were perceived. The notion of 'creative consumers' was mentioned when looking into magazines and of the *A-Z* supplement. There, it appeared that editors of *Fabulous* and *Salut les Copains* encouraged readers to appropriate the goods and their content. In the case of the *A-Z*, this took the form of cutting the magazine and folding it to create a new object. For regular issues of the magazines, the appropriation seems to have focused mostly on pictures and large, sometimes double-paged posters. They were meant to be removed from the magazines to be used on their own. In other words, magazines, understood as 'artefacts of mass culture, supposedly imposed, reified and productive only of passivity, are in fact re-worked, *bricolés* by their users in ways that are creative and dynamic'.[121] This

119 Although, as briefly mentioned previously, the BBC also conducted research on its listeners by using letters they sent. This was meant to better understand the audience's opinion on radio programmes, however, with strong limitations, according to Robert Silvey, as only literate listeners with strong opinions—whether positive or negative—were writing. Crisell, *Understanding Radio*, p. 205, and Robert Silvey, *Who's Listening? The Story of B.B.C. Audience Research* (London: George Allen & Unwin, 1974), pp. 29–31.

120 Christian Delporte, Claire Blandin and François Robinet, *Histoire de la presse en France: XXe-XXIe siècles* (Malakoff: Armand Colin, 2016), p. 211.

121 David Looseley, "Conceptualising Youth Culture in Postwar France", *Modern & Contemporary France* (15:3, 2007), pp. 261–275.

indicates a certain form of liberty to tinker with the materiality of the magazines, chosen by the editors during the production stage. It also changes the perception of the readers, who were expected to engage in an active way with the materiality of the artefacts, and thereby becoming creative consumers.[122] This is of particular interest for the discussion on the *dispositif*, as it reinforces the active role of the recipients, who were by no means passive actors.

Moving on to a more symbolic dimension, the inclusion of listeners-readers in the *dispositif* also brought them together as members of the same group or 'imagined community', to use the concept coined by Benedict Anderson. He sees it as a useful notion to define the construction of nationalism worldwide,[123] as nations are *imagined* in the sense that the individuals think of themselves as a group. Although they 'will never know most of their fellow-members', they are still a community, for they strongly see the nation as a 'deep, horizontal comradeship'.[124] Moving away from political readings of the imagined community, one can see here that such an imagined community existed among listeners of commercial radio stations and readers of their magazines. This idea was already present through the ways readers of *Salut les Copains* were included in a wider community.[125] Michele Hilmes developed the idea that the processes identified by Anderson to explain the construction of imagined communities via the newspaper reader can easily be applied to radio, and are sometimes even more powerful with this medium.[126] People reading the same newspaper each day is described as a 'ceremony replicated by

[122] According to Osgerby, this appropriation did not only occur with the studied magazines, but with other media, especially those targeting a youth market: 'but consideration is also given to youth's relationship with the media and the various ways young people make commercial texts and products 'meaningful'. Recognition is given to young people's ability to be 'active' consumers who create their own identities through their practices of consumption'. Osgerby, *Youth Media*, p. 5.

[123] Benedict Anderson, *Imagined Communities: Reflections on the Origin and Spread of Nationalism*, 3rd ed. (London: Verso, 2006), p. 6.

[124] Ibid., p. 7.

[125] This inclusion occurred on a few levels, by the fact that readers were addressed as *copains* and sections such as those for penpals allowed them to engage with each other.

[126] Michele Hilmes, "Radio and the imagined community", in Jonathan Sterne (ed.), *The Sound Studies Reader*, (Abingdon-on-Thames: Routledge, 2012), pp. 351–362.

thousands'[127] by Anderson, but the same can be said of the shared experience of thousands of listeners tuning in to the same programme at the same time.[128] Moreover, this sense of ritualisation and repetition fits well with the concept of *dispositif*, for it establishes power dynamics.

The existence of an imagined community is supported by the idea that there was a shared culture consumed by its members, furthering common norms and values, for instance, incarnated by the mythical speech described previously. In addition, the idea of a transnational imagined community was also a product of mythical speech. In this regard, the community of *copains* furthered by the intermedial venture of *Salut les Copains* is particularly revealing. Indeed, the joint efforts of radio programmes and magazines made it appear as something natural and timeless. In reality, the deconstruction of the myth reveals how this occurred. As mentioned, *Salut les Copains* held a regular discourse furthering the idea of such a community, even making it a feature of its magazine, through the regular publication of readers' letters and pen pal sections. Furthermore, the community of *copains* is presented as timeless, while in reality it is strongly intertwined with the advent of the baby boomer generation. Its members, when they became teenagers and young adults in the 1960s, made the core of the community of *copains*.[129] According to Vergnioux and Lemonnier, the rise of baby boomers as a generation and cultural group, in France,[130] can be summed up by a few factors: demographics, a better and stronger economy, more spending power for teenagers and young adults, and the designation of specific

[127] Anderson, *Imagined Communities*, p. 35.

[128] Indeed, 'perhaps more than the other technologies [...] radio imparted a feeling of connectedness, since everyone listening in knew that others were hearing the same programs at the same time'. Taylor, Katz and Grajeda, *Music, Sound, and Technology in America*, p. 4. Moreover, Crook was pondering upon the impact of radio culture on the construction of identities. Tim Crook, *The Sound Handbook* (Abingdon: Routledge, 2012), p. 151.

[129] In his study on French baby boomers, Jean-François Sirinelli shows how members of this generation, evolved from being *copains* (with the associated references to young teenagers influenced by light entertainment and juvenile mass culture) to being *camarades* (or comrades, in English, in reference to the stronger political activity of young people, especially around 1968). See Jean-François Sirinelli, 'Des 'copains' aux 'camarades'? Les baby-boomers français dans les années 1960', *Revue historique* (626:2, 2003), pp. 327–343.

[130] This also occurred in other Westner countries. See, for the British case and the development of a specific youth market. See Osgerby, *Youth Media*.

media for them,[131] among others. This shows how the baby boomers as an age group and as a generation were not timeless, but really the product of their time, and so were the *copains*.

In addition to the shared culture and mythical speech, part of the construction of the imagined community by the *dispositif* can be found in relation to space and its appropriation by members of the community. The first chapters of this book showed the importance of the broadcasting space and its representations within the *dispositif*. Previously in this chapter, the ways commercial radio could be listened to individually, 'under the bedclothes', were described.[132] Another dimension of the relationship between members of the imagined community of commercial radio and space is found in the stations' ability to gather masses. Concerts at the Villa Louvigny, such as The Grateful Dead in 1972, are an example of such an ability. The most-known example, however, is the *Nuit de la Nation* (Night of the Nation) concert, organised by *Salut les Copains* and Europe n°1 on 22nd June 1963 on the Place de la Nation in Paris, to celebrate the beginning of *Le Tour de France*, France's famous cycling race. This free event was the first large rock'n'roll concert in France. Around 150,000 young people attended,[133] hoping to see performances of popular celebrities and artists of the time—all personalities linked to *Salut les Copains*, such as Johnny Hallyday, Sylvie Vartan, and others. The concert was largely mediated afterwards for its alleged violence caused by *blousons noirs* (black jackets).[134] While police sources indicate that violence was actually quite limited, many French media, notably in the conservative press, were particularly vocal against the concert's participants, organisers, and artists. *Salut les Copains* was nicknamed *Salut les Voyous* (Hi, thugs), and a journalist from conservative newspaper *Le Figaro* even made a comparison with the NSDAP congress

[131] Alain Vergnioux and Jean-Marc Lemonnier, 'Les Adolescents des années soixantes: Salut les copains !' *Le Télémaque* (38:2, 2010), pp. 87–100.

[132] Moreover, as pointed out by Blandin, the number of young people having their own bedrooms increased throughout the 1960s, thus creating more and more individual listening spaces. Blandin, "Radio et magazine", pp. 134–142.

[133] Instead of the 30,000 to 40,000 people expected by the concert's organisers. Among the 150,000 who actually came, 80% were between 14 to 21 year-olds; in other words, roughly a fifth of the entire Parisian youth. Bodo Mrozek, *Jugend Pop Kultur: Eine transnationale Geschichte* (Berlin: Suhrkamp, 2019), p. 604.

[134] The *blousons noirs* are the French equivalent of the Greasers subculture.

in Nuremberg.[135] Other media, notably the *Salut les Copains* magazine, offered a more nuanced take on the events. Overall, the concert triggered many reactions and relaunched a debate on moral panic and the dangers of rock'n'roll on youth.[136] The event was also a stepping stone in French social history, for the youth were from then on perceived as an 'autonomous social category'.[137] This example of the *Nuit de la Nation* reveals how commercial radio could have an impact on the appropriation of spaces by its listeners, furthering the sense of an imagined community by bringing them together for a shared experience. This sense is further developed by other media—which do not belong to the *dispositif*—because they hold a series of discourses about youth, as a community and a social group; and, in this case, related to the *copains*.

Therefore, it appears that the *dispositif* of commercial radio did support the creation of an imagined community, due to the shared rituals and common culture of its members. Furthermore, members of the imagined community experienced the *dispositif* in both private (*e.g*. under the bedclothes) and public (*e.g. La Nuit de la Nation*) spaces. In the case of the intermedial *Salut les Copains*, this community was even something created by a series of discourses found in both magazines and radio programmes. This perception of those 'at the receiving end' of the *dispositif* adds another layer (as a community) to the layers mentioned previously, that is to say, their simultaneous role as listener-readers, consumers, and as measurable data.

References

Agamben, Giorgio. *Qu'est-ce qu'un dispositif* ?. Paris: Payot, 2007.
Allen, Graham. *Roland Barthes*, 2nd ed. London: Routledge, 2004.

135 Florence Tamagne, "La Nuit de la Nation. Jugendkultur, Rock'n'Roll und Moral Panics im Frankreich der sechziger Jahre", in Bodo Mrozek, Alexa Geisthövel and Jürgen Danyel (eds.), *Popgeschichte. Band 2: Zeithistorische Fallstudien, 1958–1988* (Bielefeld: Transcript, 2014), pp. 41–62.

136 Echoing, for instance, the fights against 'rocks and mods' in 1964 in Southern England, in places such as Bournemouth, Birghton, and Hastings. Mrozek, *Jugend Pop Kultur*, p. 620.

137 Florence Tamagne, "'C'mon everybody'. Rock'n'roll et identités juvéniles en France (1956–1966)", in Ludivine Bantigny and Ivan Jablonka (eds.), *Jeunesse oblige. Histoire des jeunes en France xixe-xxe*, (Paris, PUF, 2009), pp. 199–212.

Anderson, Benedict. *Imagined Communities: Reflections on the Origin and Spread of Nationalism*, 3rd ed. London: Verso, 2006.
Attali, Jacques. *Noise: The Political Economy of Music*, 12th ed. Minneapolis: University of Minneapolis Press, 1992.
Barthes, Roland. *Mythologies*, revised ed. Translation by Jonathan Cape. London: Vintage, 2009.
Blandin, Claire. "Radio et magazine: une offre plurimédia pour les jeunes des Sixties". *Le Temps des Médias* 21:2 (2013): 134–142.
Brand, Graham and Paddy Scannell. "Talk, Identity and Performance: The Tony Blackburn Show". In *Broadcast Talk*, edited by Paddy Scannell, 201–226. London: Sage, 1991.
Bodwing, Johannes. "Was macht Überherrn mit diesem Juwel?". *Saarbrücker Zeitung*, August 3rd, 2016. [https://www.saarbruecker-zeitung.de/saarland/saarlouis/ueberherrn/was-macht-ueberherrn-mit-diesem-juwel_aid-56919]
Chignell, Hugh. *Key Concepts in Radio Studies*. London: Sage, 2009.
Connor, Steven. "I Switch Off". In *Broadcasting Modernism*, edited by Cohen, Debra Ray, Michael Coyle and Jane Lewty, 274–293. Gainesville: University Press of Florida, 2009.
Crisell, Andrew. *Understanding Radio*, 2nd ed. London: Routledge, 1994.
Crook, Tim. *The Sound Handbook*. Abingdon: Routledge, 2012.
Das Buch der Ansager: Die Ständigen Rundfunkansager der Europäischen und der Grossen Amerikanischen Rundfunkgesellschaften in Wort und Bild (Berlin: Rothgiesser & Diesing A.G., 1932),
Defrain, Jean-Pierre and Jacques Boutelet. *RTL. 40 ans ensemble*. Paris: Calmann-Lévy, 2006.
Delporte, Christian, Claire Blandin and François Robinet. *Histoire de la presse en France: XXe-XXIe siècles*. Malakoff: Armand Colin, 2018.
Douglas, Susan J. *Listening In: Radio and the American Imagination*. New York: Times Books, 1999.
Dumont, Marc. "Aux origines d'une géopolitique de l'action spatiale: Michel Foucault dans les géographies françaises". *L'Espace Politique* 12:3 (2010). [https://doi.org/10.4000/espacepolitique.1744]
Eco, Umberto and Isabella Pezzini. "La Sémiologie des *Mythologies*". *Communications* 36 (1982): 19–42.
Ellis, John. *Visible Fictions. Cinema, Television, Video*. London: Routledge, 1982.
Ellis, S. L. "Transistor Radios". In *The Museum of Broadcast Communications: Encyclopedia of Radio*, edited by Christopher Sterling, 1413–1414. New York: Fitzroy Dearborn, 2004.
El Moumni, Maryam. "The Cultural Significance of Telecommunication Heritage Sites: The Case of Europe 1 Transmitter Building in Berus, Saarland". Master diss., Brandenburg University of Technology Cottbus-Senftenberg & Deakin University, 2017.

Ericson, Staffan and Kristina Riegert (eds.). *Media Houses: Architecture, Media, and the Production of Centrality*. New York: Peter Lang, 2010.
Ferrand, Franck. "Au coeur de l'émission Salut les Copains !". *Au Coeur de l'Histoire*. Europe 1. September 14th, 2012.
Fesneau, Elvina. *Le Poste à transistor à la conquête de la France. La radio nomade (1954–1970)*. Paris: INA Editions, 2014.
Fickers, Andreas. "'Neither Good, Nor Bad; Nor Neutral': The Historical *Dispositif* of Communication Technologies". In *Journalism and Technological Change: Historical Perspectives, Contemporary Trends*, edited by Martin Schreiber and Clemens Zimmermann, 30–52. Frankfurt: Campus, 2014.
Fickers, Andreas. "Die Anfänge des kommerziellen Rundfunks im Saarland". In *Medienlandschaft Saar von 1945 bis in die Gegenwart. Band 1: Medien zwischen Demokratisierung und Kontrolle (1945–1955)*, edited by Clemens Zimmermann, Rainer Hudemann and Michael Kuderna, 241–308. München: R. Oldenbourg, 2010.
Fiske, John and Henry Jenkins. *An Introduction to Communication Studies*, 3rd ed. London: Routledge, 2011.
Foucault, Michel. "The Confession of the Flesh". In *Power/Knowledge: Selected Interviews and Other Writings*, edited by Michel Foucault and Colin Gordon, 194–228. Harlow: Longman, 1980.
Foucault, Michel. *Surveiller et Punir: la naissance de la prison*. Paris: Gallimard, 1975.
Garcin, Jérôme, ed. *Nouvelles mythologies*. Paris: Seuil, 2007.
Goodale, Greg. *Sonic Persuasion: Reading Sound in the Recorded Age*. Urbana: University of Illinois Press, 2011.
Gray, Frances. "Fireside Issues: Audience, Listener, Soundscape". In *More than a Music Box: Radio Cultures and Communities in a Multi-Media World*, 2nd ed., edited by Andrew Crisell, 247–262. New York: Berghahn Books, 2006.
Grégoire, Menie, *Comme une lame de fond*. Paris: Calmann-Lévy, 2007.
Gripsrud, Jostein. *Understanding Media Culture*. London: Arnold, 2002.
Hall, Stuart. "Encoding/Decoding". In *Culture, Media, Language: Working Papers in Cultural Studies 1972–1979*, edited by Stuart Hall, Dorothy Hobson, Andrew Lowe & Paul Willis, 117–127. London: Taylor & Francis e-Library, 2005.
Harman, J. "Recording and Studio Equipment". In *The Museum of Broadcast Communications: Encyclopedia of Radio*, edited by Christopher Sterling, 1187–1193. New York: Fitzroy Dearborn, 2004.
Heijden, Tim van der. *Hybrid Histories: Technologies of Memory and the Cultural Dynamics of Home Movies, 1895–2005*. Maastricht: Universitaire Pers Maastricht, 2018.
Hendy, David. *Radio in the Global Age*. Cambridge: Polity Press, 2000.

Hilmes, Michele and Jason Loviglio, eds. *Radio Reader: Essays in the Cultural History of Radio*. New York: Routledge, 2002.
Hines, Mark. *The Story of Broadcasting House: Home of the BBC*. London: Merrell, 2008.
Jeck, Marc. "D'Villa Louvigny". In *Lieux de mémoire au Luxembourg. Erinnerungsorte in Luxemburg*, edited by Sonja Kmec, Benoît Majerus, Michel Margue & Pit Péporté, 209–214. Luxembourg: Saint-Paul, 2007.
Jehle, Anna. *Welle der Konsumgesellschaft: Radio Luxembourg in Frankreich, 1945–1975*. Göttingen: Wallstein Verlag, 2018.
Larroche, Valérie. *The Dispositif: A Concept for Information and Communication Sciences*. London: ISTE, 2019.
Legay, Richard. "RTL & Europe n°1 as Central Actors. The Importance of Mobility For Commercial Radio Stations During the Parisian Events of Mai 68". *Rundfunk & Geschichte* 3:4 (2018): 41–50.
Looseley, David. "Conceptualising Youth Culture in Postwar France", *Modern & Contemporary France* 15:3 (2007): 261–275.
Lussault, Michel. "Renouveler le dialogue", *Espace Temps* 68:70 (1998): 33–44.
Maréchal, Denis. *RTL. Histoire d'une radio populaire: De Radio Luxembourg à RTL.fr*. Paris: Nouveau Monde Éditions, 2010.
Moores, Shaun. *Interpreting Audiences: The Ethnography of Media Consumption*. London: Sage, 1993.
Mrozek, Bodo. *Jugend Pop Kultur: Eine transnationale Geschichte*. Berlin: Suhrkamp, 2019.
Nora, Pierre, ed. *Les Lieux de mémoire*, 3 vols. Paris: Gallimard, 1997.
Nora, Pierre. "Between Memory and History: 'Les Lieux de Mémoire'". *Représentations* 26 (1989): 7–24.
O'Donnell, William and Loreto Todd, *Variety in Contemporary English*. London: George Allen & Unwin, 1980.
Ong, Walter and John Hartley, *Orality and Literacy*, 3rd ed. London: Routledge, 2013.
Osgerby, Bill. *Youth Media* (London: Routledge, 2004.
Pear, Tom H. *Voice and personality*. London: Chapman and Hall, 1931.
Pires, Mat. "Les stars noires et *Salut les Copains*, 1962–1968". *Communication et langages* 111 (1997): 59–71.
Quillien, Christophe. *Nos Années Salut les Copains 1959–1976* (Paris: Flammarion, 2009) & Gilbert Jouin, *Salut les Copains*. Paris: Ipanema Éditions, 2012.
Richards, Keith and James Fox. *Life*. London: Weidenfeld, 2010.
Rikitianskaia, Maria. "A Transnational Approach to Radio Amateurism in the 1910s". In *Transnationalizing Radio Research: New Approaches to an Old Medium*, edited by Golo Föllmer and Alexander Badenoch, 133–140. Bielefeld: Transcript.

Roger, Philippe. *Roland Barthes, roman*, 2nd ed. Paris: Grasset, 1990.
Scannell, Paddy and David Cardiff. *A Social History of British Broadcasting, Volume 1: 1922–1939, Serving the Nation*. Oxford: Blackwell, 1991.
Shingler, Martin and Cindy Wieringa. *On Air: Methods and Meaning of Radio*. London: Arnold, 1998.
Silvey, Robert. *Who's Listening? The Story of B.B.C. Audience Research*. London: George Allen & Unwin, 1974.
Sirinelli, Jean-François. "Des 'copains' aux 'camarades'? Les baby-boomers français dans les années 1960". *Revue historique* 626:2 (2003): 327–343.
Smati, Nozha and Anne-Caroline Fiévet, "A la radio, la voix donne à écouter et à voir", 4th December 2017, INAglobal.fr [http://www.inaglobal.fr/radio/article/la-radio-la-voix-donne-ecouter-et-voir-10010]
Stamm, Michael. "The Sound of Print: Newspapers and the Public Promotion of Early Radio Broadcasting in the United States". In *Sound in the Age of Mechanical Reproduction*, edited by David Suisman and Susan Strasser, 221–241. Philadelphia: University of Pennsylvania Press, 2010.
Starkey, Guy. *Radio in Context*. Basingstoke: Palgrave Macmillan, 2004.
Stauff, Markus. "Materiality, Practices, Problematization: What Kind of *Dispositif* Are Media?". In *Materializing Memories: Dispositifs, Generations, Amateurs*, edited by Susan Aasman, Andreas Fickers and Joseph Wachelder, 67–83. New York: Bloomsbury, 2018.
Sterne, Jonathan. "Sounds Like the Mall of America: Programmed Music and the Architectonics of Commercial Space". *Ethnomusicology* 41:1 (1997): 22–50.
Street, Seán. *Crossing the Ether: Pre-War Public Service Radio and Commercial Competition, 1922–1945*. Eastleigh: John Libbey Publishing, 2006.
Tamagne, Florence. "La Nuit de la Nation. Jugendkultur, Rock'n'Roll und Moral Panics im Frankreich der sechziger Jahre". In *Popgeschichte. Band 2: Zeithistorische Fallstudien, 1958–1988*, edited by Bodo Mrozek, Alexa Geisthövel & Jürgen Danyel, 41–62. Bielefeld: Transcript, 2014.
Tamagne, Florence. "'C'mon everybody'. Rock'n'roll et identités juvéniles en France (1956–1966)". In *Jeunesse oblige. Histoire des jeunes en France xixe-xxe*, edited by Ludivine Bantigny and Ivan Jablonka, 199–212. Paris: PUF, 2009.
Taylor, Timothy Dean, Mark Katz and Tony Grajeda, eds. *Music, Sound, and Technology in America: A Documentary History of Early Phonograph, Cinema, and Radio*. Durham: Duke University Press, 2012.
Tinker, Chris. "Mixed Masculinities in 1960s British and French Youth Magazines". *The Journal of Popular Culture* 7:1 (2014): 84–108.
Tinker, Chris. "Shaping 1960s youth in Britain and France: Fabulous and Salut les Copains". *International Journal of Cultural Studies* 14:6 (2011): 641–657.

Vergnioux, Alain and Jean-Marc Lemonnier. "Les Adolescents des années soixantes: Salut les copains!". *Le Télémaque* 38:2 (2010): 87–100.
Zenkine, Serge."Les Indices du mythe". *Recherches & Travaux* 77 (2010): 21–32.
Zenkine, Serge. "L'Esthétique du mythe et la dialectique du signe chez Roland Barthes". *Littérature* 108 (1997): 102–124.

CHAPTER 12

Conclusion

1 Historicising the *Dispositif*

The *dispositif* can appear as a rather fixed concept shedding light on dynamics of power, however, it should be understood as 'a diagram that channels, maps, and organises movements',[1] meaning it is subject to change.[2] Even taken in the context of a smaller period—the Long Sixties—such changes in the *dispositif* are apparent, and the following pages show the flexibility of the concept. Moreover, this means that there is no intent to provide a 'model' that could explain the overall and timeless apparatus and functioning of commercial radio stations. Instead, there is a reading of this apparatus at a given time, which anchors the concept into its time through the ways developed below.

First, the collaboration with the *Fabulous 208* and *Salut les Copains* magazines represents a clear example of the aforementioned flexibility. The connections and links to this element had clear beginnings—the first publication of *Salut les Copains* in 1962 and the rebranding of *Fabulous*

[1] Markus Stauff, "Materiality, Practices, Problematization", in S. Aasman, A. Fickers and J. Wachelder (eds.), *Materializing Memories: Dispositifs, Generations, Amateurs* (New York: Bloomsbury, 2018), pp. 67–83.

[2] Shown notably by the work of Tim Van der Heijden, who analyses a century of changes. See Tim van der Heijden, *Hybrid Histories: Technologies of Memory and the Cultural Dynamics of Home Movies, 1895–2005* (Maastricht: Universitaire Pers Maastricht, 2018).

R. Legay, *The Cultural Histories of Radio Luxembourg and Europe n°1*,
https://doi.org/10.1007/978-3-031-46250-4_12

to *Fabulous 208* in 1964—and gradual endings. The end of the *Salut les Copains* radio show in 1969 is the first crucial step in the distancing of the magazine with Europe n°1, although the connections were still likely in the minds of readers. In a similar way, the collaboration between *Pilote* and Europe n°1, which was at the centre of a case study in Chapter 10, can be read as the addition of a new element to the *dispositif*, and the subsequent creation of a new transmedial product: *Le Feu de Camp du dimanche matin*. However, for some reasons, the partnership was rather short-lived, and the programme only aired for 13 episodes. This example shows how the *dispositif* was also capable of opening up to experiments with other media: some successful, like in the case of *Salut les Copains*, some less, such as *Feu de Camp*, nevertheless, they show the possibility of creating, to some extent, new interrelations within the ensemble. Then, another aspect which historicises the *dispositif* is technology, whose history accompanies changes on two ends of the infrastructure. On one hand, technological evolutions impacted transmitters, which subsequently gained more power, more reach, and improved the quality of listening. This was a crucial consideration for the stations which paid close attention to those developments, for they impacted the broadcasting space. On the other hand, radio receivers also faced technological evolutions during the period, especially with the development and growing popularity of the transistor set. Its light weight and mobility slowly introduced new practices of radio listening and new spaces, which were developed when discussing receivers. Much more could be said about the impact of these technological changes, however, it suffices to demonstrate how they could bring evolution and adaptation to the wider apparatus.

Mentioned on a few occasions throughout the book, the year 1966 represented a moment of change for the French service of Radio Luxembourg, which faced new management. This change affected the *dispositif* on multiple locations, from a rebranding of the station to RTL, marking some symbolic distancing from the Grand Duchy, to changes in the Parisian offices and studios. Another subsequent change was in the programmes, and, by extension, to the very sound of the station. By dropping a famous host such as Zappy Max and the broadcast of the long-lasting serial *La Famille Duraton* and introducing a disc-jockey such as *Le Président* Rosko (and his *Minimax* show), who made full use of the new modern studio, RTL clearly indicated a will to sound different. Moreover, this illustrates another factor of change in the *dispositif*: the competition between the various broadcasters. Part of the reason for the large-scale changes in 1966 was the decreasing share of listenership of Radio Luxembourg, notably against the direct competitor, Europe n°1. The importance of this competition on decision-making, which subsequently resulted in changes to promotion and to the *dispositif* appeared throughout the research. This was notably indicated by a document in RTL Group archives in which a staff member listened to the station and to Europe n°1 during the same day to note where the competitor was perceived as more 'modern' than its counterpart.[3] Similarly, numerous discussions during board meetings concerned survey results, here again inferring the close attention paid to this competition. It is important to keep in mind that change and evolution were not just happening, they were craved, often advanced by the commercial nature of the stations. This willingness to change is probably best illustrated by the transmitter site of Europe n°1 in the Saarland. As previously mentioned, the building in this location was designed in such a way that the main hall could welcome new technologies which did not yet exist, showing how change represented a key factor from the very beginning. Moreover, it can be argued that competition between the commercial stations brought them closer together, for many decisions made by a station in terms of programming and improvement of the transmitters were made in relation to the other station's situation. This take on competition reinforces the relevance of studying them and their *dispositif* together.

Finally, while competition between commercial broadcasters played a key role in the evolution of the *dispositif*, the relations with other media were also important. When the protests of May 1968 occurred in France, the public broadcaster of both television and radio, the ORTF, went on strike. This phenomenon led to a 'media vacuum', in which commercial stations jumped in to replace the then almost silent ORTF, to inform French listeners about the events of a particularly unstable period.[4] Therefore, for a couple of months, the purpose of the *dispositif* of commercial radio changed dramatically. Moreover, this peculiar socio-political situation might also explain the rather small impact and recognition of the joint show of RTL and Europe n°1. This common show was on its own

[3] 'Résultats d'écoute comparés', 10th January 1966, RTL Group Archives, Luxembourg.

[4] Richard Legay, "The Role of Commercial Radio Stations in the Media Vacuum of Mai 68 in Paris", *VIEW Journal of European Television History & Culture* (6:12, 2017), pp. 1–11.

a big change in the *dispositif* of both stations, which, for a night, joined hands in production, content, and reception. However, what was sold as a great moment in radio history—because of the historic competition as well as the stereophonic experiment—seems to have gone rather unnoticed. This might be explained by the new situation, a powerful media event which eclipsed the joint programme.

There is a risk, through the process undertaken here and in the previous chapter, to essentialise the *dispositif*. However, looking closely at some of its changes through the studied period is a way to avoid this pitfall, for the concept modifies interrelations, opens new ones, and closes others. Changes, adaptations, and experiments are all forms through which the *dispositif* of commercial radio stations evolved during the Long Sixties. It would not be accurate to see change as a byproduct or secondary characteristic of the topic at hand, for the stations made it central in their *dispositif*, mostly due to their commercial nature in a capitalist society. Moreover, these last pages also reinforce the importance of anchoring the book and its take on the *dispositif* within the context of the history of the 1960s.

2 Carol, Elaine, Their 10 Guineas, and the *Dispositif*

The previous chapter shed light on the various constitutive elements of the *dispositif* of commercial radio stations, and by doing so, has also deconstructed it. It is now relevant to segue back into the concept as a whole ensemble. Throughout this book, various layers (*i.e.* technical, commercial, imagined) have been used; it is, however, particularly challenging and complex to bring them together theoretically and narratively without essentialising the result. In order to give a concrete overview and a sense of the results, the following pages show, from a concrete historical example, how the *dispositif* manifested for two people in 1967. They should be read as a help to get an overview of the *dispositif*, rather than a comprehensive take on the concept. In August 1967, the magazine *Fabulous* published a short piece written by two of its readers, Carol Scurry and Elaine Montrose, who tell the story of their encounter with a celebrity in a small café in London. The text of the story is transcribed below:

'IN A SMALL CAFE. Carol Scurry and Elaine Montrose, of Gidea Park, Romford, sent us this account of their meeting with Engelbert. They win 10 gns.[5]

March 31st may have been an ordinary day for many people, but it was one of the most marvellous of our lives. For that was the day we went to see the opening of The Walkers' tour at Finsbury Park. We spent the day in London before the show, to forget worrying thoughts like, "Will Scott turn up?" or "Will Engelbert back out?" and arrived at the theatre an hour and a half early. By this time we were dying of thirst, so we decided to regain some needed energy in a small, crowded Italian cafe, near the theatre. As we sat down, staring into cups of hot coffee, dreaming of Scott, John, Gary and Engelbert, two girls came in and said in suppressed whispers, "*Please*, let us have *his* autograph'!". We thought they were referring to some local group, but still asked who they were talking about. We both did our nut when they chorused, "Engelbert HUMPERDINCK! The one in the sheepskin jacket!". After we'd picked ourselves up from the floor, we made a bee-line for him; he was so gorgeous! He joked at first that it wasn't him, so Elaine said, "I know it's you. I've seen you on tele!" We felt such nuts! He laughed, so we asked for autographs, adding we'd scream for him on the show, which must have pleased him 'cos he started signing! As we stood there, talking about anything we realised how much better looking he was in real life: lovely deep brown hair, gorgeous hazel eyes, long sideburns … mmmm. He was so nice, not at all affected with his success. Someone with him yelled from the other side of the cafe, "Hey, Gerry!" and we both said, "That's you, isn't it?" Why we said such nutty things we'll never know. We left him in peace again and both clutched his autographs, saying chokingly, "We've touched him!" You can imagine how we felt when we saw him again later on his fantastic act!.'[6]

Here; two young women met a famous singer in a café in London in 1967. How is this relevant to a discussion about the *dispositif* of commercial radio stations in the Long Sixties? This question is a fair one. After all, the word 'radio' is not even mentioned once in the article; which is actually one of the reasons why this article was picked. This absence of radio is essential, because it suggests one of the strengths of the *dispositif*: its invisibility. The whole book focused on shedding light

[5] According to the online tool of the British National Archives for currency conversion, 10 guineas in 1967 would be worth around £185 in 2017. [https://www.nationalarchives.gov.uk/currency-converter/] (last accessed 15 March 2023).

[6] *Fabulous 208*, 26th August 1967.

on the concept, its interrelations, dynamics, and constitutive elements; and, naturally, this dispelled its invisibility, which is why it is important to begin by acknowledging it.

Little is known about the main protagonists of this section: Carol and Elaine. They were friends and they live in Romford, a town near London. Keeping in mind the main readership of *Fabulous 208*, it seems fair to assume they were in their teenage years in 1967. The language used, balancing from slightly formal 'For that was the day we went to see…' to slang 'We felt such nuts!', seems to betray their age. Carol and Elaine were likely regular readers of the magazine, as they knew that editors published texts sent their way. Once it is established that the young women are familiar with *Fabulous 208*, the *dispositif* starts to appear, for the magazine bears its connection to the English service of Radio Luxembourg in its very title; in this case, the '208' refers to the station's wavelength. Additionally, any reader skimming through an issue of the magazine would come across programme listings for the station, and articles about, or even written by, some of its disc-jockeys. As readers of *Fabulous 208*, it is therefore likely that Carol and Elaine were familiar with Radio Luxembourg's programmes and disc-jockeys, even if they were not listeners. And the same thing happened across the Channel. Readers of *Salut les Copains* would find an abundance of references to the radio show in their magazines. This familiarity is one of the strengths of the *dispositif*; radio stations were present in magazines even for readers who would not have cared for them. According to letters they sent to the editors, some readers of both magazines were living in countries way outside of the reach of the stations. In spite of the distance, they would have known of Radio Luxembourg and Europe n°1, their programmes, and their hosts.

Such intermedial dynamics occurred both ways, as listeners would have heard about magazines during programmes. Daniel Filipacchi, for instance, made sure of that. When he presented *Salut les Copains*, he referred on a few occasions to the magazine. Claude François did the same thing when he casually skimmed through an issue of *Salut les Copains* on air, wherein many pages featured him. Filipacchi also mentioned at times—both intentionally and as a slip of the tongue—*Mademoiselle Âge Tendre*, another magazine he edited, which shows the variety of actors involved beyond those studied in this book. Alternatively, commercial stations could be listened to in Scandinavia and across the Iron Curtain, where listeners may have heard the same records as Carol and Elaine.

This means that the way through which consumers 'enter' the *dispositif* does not matter; a reader would read about the station, while a listener would hear about the magazine. The reason why is simple; by having various media promote each other within the *dispositif*, they could support sales and shares of listenerships, and, by extension, increase revenue streams. The commercial dimension of the *dispositif* cannot be described as invisible; a plethora of adverts—the most tangible expression of the commercial nature—are found in the programmes and in magazines. These adverts, however, are casted, moulded, into the content produced by both media, to the extent commercialism simply becomes a regular part of them. Many commodities are promoted through adverts, from items like record players and shaving cream, to more immaterial products, such as songs and celebrities; or, at least, images of celebrities. Through advertisement, readers of *Fabulous 208* like Carol and Elaine were confronted with the multiple interrelations of the *dispositif* within the wider consumerist popular culture.

In the brief yet salient article, such interrelations with popular culture are already present. The young women put a lot of importance on live concerts for example, as the discussed gig was truly a big event for them. They describe it as 'fantastic' and as 'one of the most marvellous [days] of [their] lives'. Indeed, these events are not just about music, they are also socially important, as illustrated previously in this chapter by the example of the *Nuit de la Nation*. Carol and Elaine even spent the whole day in London with 'worrying thoughts', due to possible tensions between 'Scott, John and Gary', the siblings of the American band the Walker Brothers, and their guest, Engelbert Humperdinck. This obviously points to a wider culture of stardom, wherein celebrities' personal lives play a large part, as illustrated by Carol and Elaine's worries. How they knew about these tensions is unclear. However, it is not impossible that they read about it in a magazine like *Fabulous 208*. The two young women recognised Engelbert, and so did others in the café, showing the importance of shared norms, values, and knowledge in a social group or imagined community. This event also hints at another aspect of stardom: the artist's face is well-known. Indeed, several photos of artists could be found in each issue of youth magazines such as *Fabulous 208* and *Salut les Copains* (and in many others). Although Elaine explained she was positive it was him not thanks to a magazine, but because she 'saw [him] on tele!'. By saying so, she conjures another medium—television—into the set of complex interrelations discussed in this section. When the young women wrote something with sexual undertones such as 'gorgeous hazel eyes, long sideburns … mmmm' when describing Engelbert, they echo an intermedial feature of the *dispositif*: its irreverence, mentioned on several occasions in this book. This goes to show how readers could, to some extent, appropriate some of the codes and features of both media. Radio, television, and magazines all contributed to stardom overall, and one aspect of this phenomena is the representation of celebrities as both unique individuals and as common people. Carol and Elaine's brief anecdote published in a widely read magazine contributes perfectly to this representation. Indeed, while spotting Engelbert created some agitation (screams, mostly), Carol and Elaine thought that 'he was so nice, not at all affected with his success', generously reinforcing this duality of celebrities.

Clearly illustrated by this example, *Fabulous* and *Salut les Copains* both published pieces written by their readers. Stories such as the one mentioned here, but also letters and profiles of those interested in finding pen pals. The magazines interacted with their readership through many ways, as did the stations thanks to phone-ins. Among all the potential examples, Carol and Elaine's story was picked for this section because the celebrity they met was none other than Engelbert Humperdinck. The same Engelbert who accompanied a team from *Fabulous* to visit the Grand Duchy and the studios of Radio Luxembourg. As detailed at length in a previous chapter, the photo-reportage was a celebration of the 'link-up' between *Fabulous* and Radio Luxembourg. In the photo-reportage, readers could see Engelbert being photographed all around Luxembourg, which was portrayed as a foreign and exotic place; at least, from a British perspective. The singer was also photographed in the studios of the Villa Louvigny, where disc-jockeys interviewed him on air. Logically, this means that listeners heard his voice. It does not seem far-fetched to assume the station also played some of his records around that time, considering he was picked to visit the studios and was interviewed on air. The photo-reportage and the interview reminded both readers and listeners—to whom Carol and Elaine might have belonged—of the transnational anchoring of the station. This complex transnational relationship between offices, studios, and transmitters—all at the heart of the photo-reportage—also applied to Europe n°1 and the French service of Radio Luxembourg. Anyone tuning in to one of the commercial stations—the *radios périphériques*—was actually confronted to this dynamic. Indeed, the broadcasting side of the *dispositif* was alluded to in both media, and so was the reception side. Both accidental and deliberate examples abound in the soundscape of commercial radio stations. Ranging from an impromptu phone call, to a radio host on air, to clear instructions on how to listen to the joint programme by RTL and Europe n°1 by way of interference from another station, listeners were regularly confronted to the materiality of the radio experience. The aforementioned act of tuning in represents a key aspect of the *dispositif* for it connects listeners to many commodities produced by the commercial radio stations. Moreover, it brings to the centre the importance of communication technologies linked to radio broadcasting. Transmitters, receivers, transistors, and even radio dials all contribute to the existence of the *dispositif*.

This echoes a point made previously: it does not matter from where the *dispositif* is conceived, every dimension and layer is found within the others, and all constitutive elements reference each other. This is why the article above was picked. Although it did not mention radio directly, the medium still found its way into the discussion. Even from a rather mundane article, the threads of the *dispositif* can be unwoven.

3 Final Remarks

The previous chapter offered an analysis and an overview of the *dispositif* of commercial radio stations by building on the results obtained throughout the previous chapters. Therefore, they have proven the existence of such a *dispositif* in the Long Sixties, which was the central hypothesis. This book revealed its functioning thanks to rich and varied source material, and through three main perspectives, (1) its appropriation of a transnational broadcasting space, (2) its use of a rich soundscape which differentiated from a public model, (3) its reliance on entanglements with other media, mostly youth magazines, and, to a lesser extent, on *bandes dessinées*.

While institutional archives and youth magazines have been used many times by historians, other sources are still almost untapped resources. This is the case for instance of private collectors, who hold a large amount of historical recordings. Other unusual sources, such as maps and recordings from the archives of Europe 1, are unfortunately more difficult to access.[7]

[7] The author was only able to access them after lengthy negotiations, a sheer amount of luck, and the support of people in Luxembourg, Paris and Saarbrücken.

This book contributes to shed some light on a few of the blind spots mentioned at the beginning. First, the research undertaken here has helped further the understanding of the history of Europe n°1 and Radio Luxembourg in the 1960s, notably by looking at them together, rather than separately, as it is usually the case. This has led to a change of perspective throughout the research, which, subsequently, has brought a better knowledge of the European broadcasting landscape in the Long Sixties.[8] By not treating commercial stations as marginal, nor by looking at them separately, this study has contributed to reinforce the importance of commercial broadcasters as a different *dispositif*, which was built in opposition to public radio, whose history still dominates the historiography. The history of popular culture in Western Europe in the 1960s as a whole benefits from the research undertaken in these pages, which has closely studied some of the most popular media and products (*e.g.* radio shows, youth magazines, and, to some extent, *bandes dessinées*) of the time. Additionally, focusing on commercial radio naturally leads to embracing a transnational perspective, and, by doing so, this book follows in the footsteps of several media scholars who have called for such a transnational turn.[9] The opening up of a national framework was not the only academic call this work has answered. By focusing on the 'plurimedia offer' of commercial radio stations in the Long Sixties, this study has contributed to the necessary take to look at media history beyond a single-medium lens. The intermedial and intertextual analysis of both magazines and radio programmes is an example of how this call for cross-media perspectives can be applied.

Naturally, by choosing to go for transnational and especially intermedial perspectives, this book has embraced a rather broad understanding of what radio is. Following Kate Lacey's reflections, the 'very idea of radio' here is undoubtedly more than just broadcasting.[10] While the

[8] Echoing, for instance, the works done by Suzanne Lommers and Christoph Classen. See Suzanne Lommers, *Europe—On Air. Interwar Projects for Radio Broadcasting* (Amsterdam: Amsterdam University Press, 2012) and Christoph Classen (ed.), *Transnational Broadcasting in Europe 1945–1990* (Frankfurt am Main: Peter Lang, 2016).

[9] See Golo Föllmer and Alexander Badenoch (eds.), *Transnationalizing Radio Research: New Approaches to an Old Medium* (Bielefeld: Transcript, 2018).

[10] Indeed, 'the singular word, radio, is called upon to describe any number of different things—material, virtual, institutional, aesthetic, experiential'. Kate Lacey, "Up in the Air? The Matter of Radio Studies", *Radio Journal: International Studies in Broadcast & Audio Media* (16:2, 2018), pp. 109–126.

production and content of radio have been a strong focus, the use of various dimensions (*e.g.* material, commercial, perceived) already opened up the definition of radio. This definition is even further developed by the input of intermediality, incarnated in the use of varied source material throughout the research. As such, this work has avoided the danger 'to isolate radio, to separate it off from its rightful connections with the more established currents in media and cultural studies, let alone disciplines further afield'.[11] Radio, throughout this work, has been understood as a vessel for commercial commodities, programmes, magazines, sonic icons, and several other elements. And this was possible thanks to the use of the concept of *dispositif*, which has allowed the research to avoid essentialisation, and embrace the porous edges of radio.[12] Furthermore, by focusing on the *dispositif*, this book has decentred the underlying 'idea' of radio, meaning that the specificities of the medium have been present, but have in no way acted as limits to the discussion and the analysis, which has often reached out to other media and other fields of research.

This take on the idea of radio is perceived, in the author's mind, as the central and most tangible contribution of this book to the wider field of radio history. This contribution is understood as twofold. First, the theoretical take on the medium and on its history as developed above. Second, through the empirical application of said discussion. Indeed, this book has applied several methodologies and developed a series of examples illustrating this richness of the idea of radio, which could be used elsewhere. The study of the sound could easily be applied elsewhere, while the intermedial and intertextual analysis of the chapters on entanglements have only been looking at a small portion of the entire interconnections between media. Commercial radio was entangled with other magazines than *Fabulous 208* and *Salut les Copains*, and with other media as well, such as television, showing that there is still a lot to investigate. The case study of *Feu de Camp* is likely another crucial contribution to media history overall. While the findings of the case study are compelling, it is the method used which is likely more useful to the wider field. Indeed, it has shown that collaboration between historians of various disciplines

[11] Kate Lacey, "Ten Years of Radio Studies: The Very Idea", *The Radio Journal: International Studies in Broadcast & Audio Media* (6:1, 2008), pp. 21–32.

[12] As mentioned by Carolyn Marvin: 'media are not fixed objects: they have no natural edges'. Carolyn Marvin, *When Old Technologies Were New* (New York: Oxford University Press, 1988), p. 8.

bears fruit and is even necessary to shed light on some blind spots of media history. The same case study but conducted only by a radio historian or only a comics historian would have missed on many aspects.

To conclude these final remarks, the author would like to insist on the importance and pleasure of working with a rich variety of source material. Not only does it help to open up the 'idea' of radio to capture it in all its richness, it is also an enjoyable endeavour. There is still a lot to unearth about the cultural history of commercial radio stations. As they are a perfect soil for a wide range of further analyses, whether cultural, intermedial, or transnational, it is expected that they will be at the centre of attention of many more works. The author also hopes that this book will be helpful to anyone interested in similar perspectives and topics, either through the findings or the methods used throughout this work.

References

Classen, Christoph, ed. *Transnational Broadcasting in Europe 1945–1990*. Frankfurt am Main: Peter Lang, 2016.

Föllmer, Golo and Alexander Badenoch, eds. *Transnationalizing Radio Research: New approaches to an old medium*. Bielefeld: Transcript, 2018.

Heijden, Tim van der. *Hybrid Histories: Technologies of Memory and the Cultural Dynamics of Home Movies, 1895–2005*. Maastricht: Universitaire Pers Maastricht, 2018.

Lacey, Kate. "Up in the Air? The Matter of Radio Studies". *The Radio Journal: International Studies in Broadcast & Audio Media* 16:2 (2018): 109–126.

Lacey, Kate. "Ten Years of Radio Studies: The Very Idea". *The Radio Journal: International Studies in Broadcast & Audio Media* 6:1 (2008): 21–32.

Legay, Richard. "The Role of Commercial Radio Stations in the Media Vacuum of Mai 68 in Paris". *VIEW Journal of European Television History & Culture* 6:12 (2017): 1–11.

Lommers, Suzanne. *Europe - On Air. Interwar Projects for Radio Broadcasting*. Amsterdam: Amsterdam University Press, 2012.

Marvin, Carolyn. *When Old Technologies Were New*. New York: Oxford University Press, 1988.

Stauff, Markus. "Materiality, Practices, Problematization: What Kind of *Dispositif* Are Media?". In *Materializing Memories: Dispositifs, Generations, Amateurs*, edited by Susan Aasman, Andreas Fickers and Joseph Wachelder, 67–83. New York: Bloomsbury, 2018.

(continued)

Europe n°1

Name	*Length*	*Language*	*Date*	*Archives*
1er indicatif	00:00:37	FR/DE/EN	1955	R. Biesen
Pour ceux qui aiment le jazz	00:20:30	FR	1955	Europe 1
Le Café de l'Europe	00:02:17	FR	01/04/1955	Europe 1
Entretien avec Salvador Dali	00:04:13	FR	02/05/1955	Europe 1
Signé Furax épisode 5	00:08:34	FR	26/10/1956	Europe 1
Ma vie a son secret (extrait)	00:02:41	FR	06/01/1957	Europe 1
Musicorama	01:01:05	FR	09/01/1957	Europe 1
Pour ceux qui aiment le jazz	00:19:17	FR	29/03/1957	Europe 1
Signé Furax épisode 247	00:08:48	FR	29/06/1957	Europe 1
Un million cash	00:04:36	FR	1957	Europe 1
Europe Matin	00:17:14	FR	26/05/1958	Europe 1
Europe Midi	00:02:21	FR	25/03/1958	Europe 1
Musicorama (1ère partie)	00:29:04	FR	01/05/1958	Europe 1
Naissance d'une étoile	00:02:25	FR	??/10/1958	Europe 1
Tour du Monde en 98 jours	00:39:18	FR	30/10/1958	Europe 1
Vous êtes formidables	00:03:53	FR	25/3/1958	Europe 1
Journal parlé d'Europe 1	00:09:24	FR	18/02/1959	INA
Le Coq de la Chanson Française	00:48:13	FR	1959	Europe 1
Le roman des vedettes	00:07:19	FR	10/12/1959	Europe 1
Le Sourire de Francis Blanche	00:02:54	FR	1960	Europe 1
Bons baisers et à bientôt	00:02:20	FR	28/08/1960	Europe 1
Le Petit pensionnaire	N/A	FR	24/12/1960	Europe 1
Vive le music-hall	00:32:58	FR	25/08/1060	Europe 1
Ah ah ah	00:10:12	FR	1961	Europe 1
Canular téléphonique	00:01:24	FR	1961	Europe 1
Dans le vent (Claude François)	00:53:55	FR	24/01/1961	Europe 1
Midi à 14 h	00:44:28	FR	26/03/1961	Europe 1
Un Millionnaire au bout du fil	00:24:46	FR	24/09/1961	Europe 1
Vous êtes formidables	00:13:05	FR	24/12/1961	Europe 1
Journaux parlés	00:23:13	FR	27/08/1962	INA
Messe à Milan, Cardinal Montini	00:04:13	FR	Noël 1962	R. Biesen
Le Magazine du spectacle	00:12:51	FR	04/05/1962	Europe 1
Equipe n°1	00:55:51	FR	20/05/1962	Europe 1
M. Monroe chante	00:00:59	EN	20/05/1962	Europe 1
Noël autour du Monde	00:12:55	FR	24/12/1962	Europe 1
Europe soir	00:22:03	FR	31/07/1963	INA
Chronique de Claude Terrien	00:09:51	FR	03/12/1963	INA
Chronique de Claude Terrien	00:08:30	FR	04/12/1963	INA

(continued)

(continued)

Europe n°1

Name	*Length*	*Language*	*Date*	*Archives*
Journal parlé	00:52:36	FR	17/07/1963	INA
Lettres d'amour	00:08:38	FR	29/01/1963	Europe 1
Dans le vent (Dalida)	01:48:46	FR	17/12/1963	Europe 1
Salut les Copains (extrait)	00:02:01	FR	27/08/1964	Europe 1
Dans le vent (E. Macias & H. Aufray)	00:56:38	FR	20/05/1964	Europe 1
Gardez le sourire	00:04:08	FR	1965	Europe 1
Bonjour M. le Maire (Malbuisson)	00:10:15	FR	20/03/1965	Europe 1
Salut les Copains (C. François)	00:39:21	FR	00/04/1965	Europe 1
Salut les Copains (1ère partie)	01:00:04	FR	21/05/1965	Europe 1
Salut les Copains (2ème partie)	00:45:51	FR	21/05/1965	Europe 1
Salut les Copains (extrait)	00:01:12	FR	01/11/1966	Europe 1
Salut les Copains	00:01:08	FR	00/00/1966	Europe 1
Salut les Copains (extraits)	00:31:00	FR	00/00/1966	Europe 1
Salut les Copains	00:01:49	FR	08/07/1966	Europe 1
Salut les Copains	00:00:29	FR	1967	Europe 1
Salut les Copains (J. Hendrix)	00:01:58	FR	29/01/1968	Europe 1
Bonjour M. le Maire (Bouzonville)	00:10:00	FR	1969	R. Biesen
Fou-rire pendant une publicité	00:01:36	FR	1969	Europe 1
Fou-rire pendant un journal parlé	00:03:04	FR	1969	Europe 1
Cafouillage sur la radioguidage	00:01:21	FR	1969	Europe 1
Le feu de camp du dimanche matin	00:46:39	FR	02/11/1969	Europe 1
Le feu de camp du dimanche matin	01:22:11	FR	23/11/1969	Europe 1
Hubert	00:31:34	FR	1970	Europe 1
Feuilleton des animateurs d'Europe 1	00:19:19	FR	09/11/1970	Europe 1
H. Aufray sur Europe 1 avec Vonny (1ère partie)	00:57:22	FR	02/07/1971	Europe 1
H. Aufray sur Europe 1 avec Vonny (2eme partie)	00:14:25	FR	02/07/1971	Europe 1
Jackpot (extrait)	00:00:38	FR	05/01/1972	Europe 1
Jackpot (extrait)	00:01:08	FR	23/06/1972	Europe 1
Requiem pour une galaxie	00:05:55	FR	1972	Europe 1
Il y a sûrement quelque chose à faire	00:24:35	FR	25/12/1973	Europe 1
Mozik	01:24:41	FR	5/10/1973	Europe 1

(continued)

(continued)

Europe n°1

Name	*Length*	*Language*	*Date*	*Archives*
Le rendez-vous d'Europe Soir	N/A	FR	22/04/1974	INA
Bonjour M. le Maire (Merten)	00:09:42	FR	années 60	R. Biesen
Bonjour M. le Maire (Remering)	00:10:01	FR	1965/66	R. Biesen
Indicatif	00:01:36	N/A	?	R. Biesen

Radio Luxembourg (English service)

Name	*Length*	*Language*	*Date*	*Archives*
Fragments	00:10:16	EN	1963	Personal
Your record Show - Brian Matthew	00:25:44	EN	1963	Personal
Barry Alldis	00:56:18	EN	01/08/1965	Personal
Keith Fordyce	00:37:23	EN	1966	Personal
Barry Alldis	00:56:12	EN	26/06/1966	Personal
Jack Jackson	00:27:20	EN	10/07/1966	Personal
Christmas Show (Jimmy Saville)	00:30:32	EN	24/12/1966	Personal
Paul Kaye	00:30:29	EN	01/02/1968	Personal
Noel Edmonds	00:21:09	EN	19/10/1968	Personal
Paul Burnett	00:58:25	EN	09/06/1969	Personal
David 'Kid' Jensen	00:59:10	EN	13/06/1969	Personal
Paul Burnett	00:32:19	EN	13/06/1969	Personal
David 'Kid' Jensen & Ringo Starr	00:16:49	EN	26/09/1969	Personal
Tony McArthur & John Lennon	00:22:59	EN	27/09/1969	Personal
David 'Kid' Jensen	00:54:16	EN	14/11/1969	Personal
Tony Prince Show	00:31:37	EN	15/06/1970	Personal
Bob Stewart	00:45:43	EN	21/06/1970	Personal
Tony Prince Show	00:59:09	EN	17/08/1970	Personal
Tony Prince Show	00:23:53	EN	17/08/1970	Personal
Esso show - Kenny Everett	00:29:07	EN	13/11/1970	Personal
Christmas Day Kenny Everett	00:24:18	EN	25/12/1970	Personal
David 'Kid' Jensen	00:45:22	EN	08/03/1971	Personal
Paul Burnett Show	01:00:20	EN	19/08/1971	Personal
Paul Burnett Show	00:58:44	EN	19/08/1971	Personal
Bob Stewart	01:00:12	EN	19/08/1971	Personal
Bob Stewart	01:00:14	EN	19/08/1971	Personal
David 'Kid' Jensen & Dave Christian	00:58:33	EN	20/08/1971	Personal
Romeo	00:59:53	NEE/EN	28/08/1971	Personal
Romeo & Mark Wesley	00:59:39	NEE/EN	28/08/1971	Personal
Mark Wesley Show	00:28:49	EN	28/08/1971	Personal

(continued)

(continued)

Radio Luxembourg (English service)

Name	*Length*	*Language*	*Date*	*Archives*
Peter Koelewijn	00:34:58	NEE	02/09/1971	Personal
Felix Meurders	01:03:39	NEE	02/09/1971	Personal
Felix Meurders, Paul Burnett	01:03:38	NEE/EN	02/09/1971	Personal
Paul Burnett Show	00:59:27	EN	19/09/1971	Personal
Paul Burnett, David 'Kid' Jensen	00:59:33	EN	20/09/1971	Personal
Elvis Special - Tony Prince, Mark Wesley	00:42:46	EN	1972	Personal
Bob Stewart	00:37:39	EN	27/01/1972	Personal
Dave Christian	01:13:27	EN	21/08/1972	Personal
Mark Wesley Show	01:02:47	EN	01/09/1972	Personal
Emperor Rosko	01:00:19	EN	01/09/1972	Personal
Emperor Rosko	00:52:56	EN	01/09/1972	Personal
Bob Stewart	01:00:14	EN	02/09/1972	Personal
Tony Prince Show	00:59:43	EN	23/09/1972	Personal

Mixed programmes

Name	*Length*	*Language*	*Date*	*Archives*
Journaux parlés de Radio Luxembourg et d'Europe n°1	N/A	FR	28/08/1962	INA
Journaux parlés de Radio Luxembourg et d'Europe n°1	01:12:09	FR	02/09/1962	INA
Journaux parlés de Radio Luxembourg et d'Europe n°1	N/A	FR	01/07/1963	INA
Journaux parlés de Radio Luxembourg et d'Europe n°1	01:02:50	FR	07/07/1963	INA
Journaux parlés de Radio Luxembourg et d'Europe n°1	01:07:07	FR	10/07/1963	INA
Journaux parlés de Radio Luxembourg et d'Europe n°1	00:57:44	FR	13/07/1963	INA
Journaux parlés de Radio Luxembourg et d'Europe n°1	00:23:12	FR	14/07/1963	INA
Emission commune RTL et Europe n°1 avec J. Yanne & F. Blanche	02:31:58	FR	08/05/1968	R. Biesen

Index

R. Legay, *The Cultural Histories of Radio Luxembourg and Europe n°1*, https://doi.org/10.1007/978-3-031-46250-4